An Introduction to Cosmology

Jeremy Bernstein

Prentice Hall
Englewood Cliffs, N.J. 07632

Library of Congress Cataloging-in-Publication Data

Bernstein, Jeremy
An introduction to cosmology/Jeremy Bernstein.
p. cm.
Includes bibliographical references and index.
ISBN 0-13-110504-3
1. Cosmology. I. Title.
QB981.B49 1995
523.1—dc20 94-14493
CIP

Acquisitions Editor: Ray Henderson
Project Manager: Carol Bollati
Cover Designer: Karen Salzbach
Manufacturing Buyer: Trudy Pisciotti
Text Composition: BPC Digital Techset
Text/Cover Printer: Hamilton
Art Studio: Vantage Art, Inc.

Printed in the United States of America

10 9 8 7 6 5 4 3 2 1

ISBN 0-13-110504-3

ISBN 0-13-110504-3

90000

9 780131 105041

Prentice-Hall International (UK) Limited, *London*
Prentice-Hall of Australia Pty, Limited, *Sydney*
Prentice-Hall Canada, Inc., *Toronto*
Prentice-Hall Hispanoamericana, S. A., *Mexico*
Prentice-Hall of India Private Limited, *New Delhi*
Prentice-Hall of Japan, Inc., *Tokyo*
Simon & Schuster Asia Pte. Ltd., *Singapore*
Editora Prentice-Hall do Brasil, Ltda., *Rio de Janeiro*

Contents

Acknowledgments

In addition to the students in my course who have helped me by the questions that they have asked, I am grateful to several friends and colleagues. I would like to cite E. Abrahams, J. Anderson, L. Brown, G. Feinberg, P. Kaus, M. Turner, P. J. E. Peebles, B. Ratra, and D. Wilkinson for many helpful conversations and critical remarks. I would especially like to thank S. Dodelson for his careful reading of the manuscript. I would also like to acknowledge the hospitality of the Aspen Center for Physics where much of the work on this book was done. In addition I would like to thank Carol Bollati and Ray Henderson of the Prentice Hall Company for their help in taking the project from a manuscript to a book.

Introduction and Plan of the Book

We begin by presenting a brief historical introduction to our subject—cosmology. All the races of humankind seem to have shared a common desire to account for their origins, and indeed the origins of the universe at large. (Another thing which appears to be transcultural is the brewing of alcoholic beverages.) It is not surprising that the early prescientific cosmologies tell us more about the societies that produced them than about the universe. Take, as an example, the creation passage from the Bihandaranyaka Upanishad—the Sanskrit epoch—that begins, "In the beginning nothing at all existed here. This whole world was enveloped by Death—by hunger. For what is Death but hunger? And Death bethought himself, 'Would that I had a self!' He roamed around, offering praise: and from him, as he offered praise, water was born." What a terrifying and revealing question about a society—"For what is Death but hunger?" Many of the early cosmologies depicted the universe in the shape of an egg—a homely rural image. "In the beginning of creation," begins the Book of Genesis, "when God made heaven and earth, the earth was without form and void, with darkness over the face of the abyss, and a mighty wind that swept over the surface of the waters. God said, 'Let there be light' and there was light. . . . He called the light day and the darkness night." One of the things we shall learn later in this book is that if the hot Big Bang model of cosmology is right, it took about a 100,000 yr after the Big Bang for the primal gamma rays to cool down enough so that they would appear as visible light.

In contemplating the origins of the universe, both the notion that the universe had a beginning in time and the notion that it didn't have a beginning in time seem equally implausible. But since most things with which we are familiar seem to have an origin in time, it was natural in the early cosmologies to assume that the universe as a whole was created at a definite time—usually by a creator. An exception to this prejudice in the ancient world was the view of Aristotle who promulgated the notion that the universe had no beginning but is and always was. His contention came into conflict with the major non-Asiatic post-Aristotelian religions in which creation as the act of a creator is a central part of the doctrine. A precise time of the Creation was often proposed. Indeed, in 1654, the Irish divine James Ussher, after a lifetime of studying biblical chronology, announced that the Creation had occurred on Tuesday, October 26, 4004 B.C. at 9 A.M.—presumably Greenwich mean time. The proponents of the presently accepted Big Bang theory, of which this author is one, are only willing to say that the universe is probably at least 10 billion years old and at most 15 billion years old. When it comes to precision, we cannot compete with Bishop Ussher. Newton, who spent much more of his time on alchemy and biblical dating—the sort of thing that Ussher was doing—than he did on what we would call science, did not, as far as I know, give a date for the Creation. But he did say that the end of the world could not come before the year 2060—just as well for the readers of this book.

Nonetheless, it is to Newton that we owe, like so many other things, the first scientific (in our sense) statement about cosmology. Newton raised a question which in different guises keeps coming back into the subject. Newton imagined a universe that was finite in space. Implicitly he assumed that it was sufficiently cold so that only gravitational attraction played a role—no pressure. He asked what would happen if there was any clumping of matter in this universe. He realized that because of the universal nature of gravitational attraction, such an effect would tend to run away. The clump would attract more matter, which would in turn attract even more matter until everything collapsed into a lump, contrary to what we observe. Newton understood that a way out of this dilemma was to make the universe spatially infinite so that there would be no "place" for the matter to collect. As he wrote to a contemporary, Richard Bentley, "But if matter was evenly disposed throughout an infinite space it would never convene into one mass; but some of it would convene into one mass and some into another, so as to make an infinite number of great masses scattered at great distances from one another throughout all that infinite space." This description foreshadows the modern theory of galaxy formation.

By the nineteenth century, a substantial body of geological evidence had accumulated indicating that Bishop Ussher's Creation date, as appealing as it was, had to be wrong. This was a period during which the modern science of thermodynamics was also created. One of its major architects, William Thompson (Lord Kelvin), assuming that the earth began as a hot, molten mass—and that nothing intervened in the cooling process—reasoned that the age of the earth lay somewhere between 20 million and 400 million years. By similar arguments he attributed to the sun an age of about the same order of magnitude. This was an argument—an important *scientific* argument—that was used against Darwin's theory of evolution which seemed to require countless millennia for its unfolding. By the end of the century radioactivity had been discovered, and it was realized that this source of energy could keep the earth from simply cooling thermodynamically. In this way the age of the earth was extended to some billions of years—$4\frac{1}{2}$ billion is the present best estimate—in comfortable agreement with the requirements of the theory of evolution.

If one wants to assign a birth year to what we would call modern cosmology—the subject matter of this book—1917 is as good as any. That was the year that Einstein published a paper he called "Cosmological Considerations on the General Theory of Relativity." It is a remarkable paper both for what it revealed and for what it might have revealed but didn't. Einstein began by making two simplifying assumptions. (We leave it as an exercise for the reader to show that the second actually implies the first, but not vice versa.) He assumed that the universe, in an average sense, is both homogeneous and isotropic. By "homogeneous" we mean that each observer, wherever located, will report the same set of observations, for example, the value of the matter density in his or her location. This does not mean that a given observer will necessarily see the same matter distribution in every direction. This is a separate assumption—isotropy. Of course, if I look out the window of the room in which I am writing this and compare what I see to the view from someone else's window across the city, both of these assumptions are clearly false. But in cosmology we are talking about averaging over billions upon billions of galaxies. In that sense, the assumptions begin to seem reasonable. They make our equations—beginning with those in Einstein's paper—tractable and allow us to draw definite conclusions and make predictions that can be either verified or falsified.

The next assumption Einstein made reflected the limited amount of observational knowledge that was then available to astronomers. It was not until a decade later—largely through the work of the American astronomer Edwin Hubble, who also discovered that the universe is expanding—that it was proven that there are astronomical objects (galaxies) that are actually outside our own Milky Way galaxy. For Einstein, all the astronomical data appeared to be consistent with a static, unchanging universe. It is true that there were small "proper" motions of stars within our galaxy, but there did not seem to be any secular motion of the universe as a whole. But this appeared to conflict with Einstein's own theory of relativity and gravitation. Here was just Newton's dilemma back in a new guise, namely, that an ensemble of gravitating matter is inherently unstable. In dealing with this, Newton made the universe infinite, something that apparently did not appeal to Einstein. Rather he changed his law of gravitation. It turns out that one can add an additional interaction to Einstein's equations without modifying their basic underlying symmetry. This introduces a new force which can be made suitably repulsive so as to counteract the gravitational force. By choosing the constants just right, one can arrange things so that the universe is stationary. Many years later, after it had been demonstrated that the universe is not stationary, Einstein referred to the addition of this so-called cosmological constant term as his greatest scientific "blunder." If he had stuck to his original theory, he might have actually predicted the very expansion of the universe that Hubble discovered a decade later.

There were, however, two theoreticians who did examine the possibilities of an expanding universe before its discovery. In 1917, the same year that Einstein published his paper, the Dutch astronomer Willem de Sitter produced an odd variant of it in which there was a cosmological constant but no matter. He discovered that in such a universe—so-called de Sitter space—the expansion would grow exponentially in time. This seems very unphysical. It is certainly not, as far as we know, true of the present universe, but, as we shall see, there may have been an epoch in the very early universe where these conditions obtained—something that is now referred to as inflation. It is interesting to note that when Hubble found the evidence (we shall discuss it later in the book) that the universe seemed to be expanding, he mentioned the work of de Sitter which he thought he might be confirming. Taken at face value, without a theory, Hubble's original data are not entirely convincing. He felt, it appears, the need for a theoretical model to lend them credence.

The second individual who saw the possibility of an expanding universe, but within the context of Einstein's original unmutilated theory, was a Russian polymath by the name of Aleksandr Alexandrovich Friedmann. Friedmann was not actually a physicist, but he had become—as had so many people—smitten by the general theory of relativity after it had been confirmed in 1919. It was shown that starlight passing close to the sun is bent by gravitation by an amount predicted by the theory. This was the first experimental indication that space itself might be "curved." Friedmann was a mathematician and meteorologist who by the year of Einstein's paper—1917—had somehow become the director of the first Russian factory devoted to the manufacture of aviation instruments. In his spare time, one would suppose, Friedmann discovered that Einstein's original equations had beautiful solutions in which the universe expanded indefinitely in a stately way and that it had others in which the universe stopped expanding after a finite time and recollapsed. What it actually does depends on the distribution of energy. Since in relativity mass and energy are related by Einstein's equation $E = mc^2$, the distribution of energy is what affects the gravitational properties of the expansion. We shall study Friedmann's solutions in considerable detail, and we shall examine our present state of knowledge about the distribution of energy.

It took some years—at least until 1929 when Hubble published his discovery—before Friedmann's work was appreciated. In fact, Einstein at first said that it was wrong but then retracted his claim when he realized that it wasn't. Unfortunately, Friedmann died in 1925 (at age 37)—before Hubble did his work. The two men most closely associated with the post-Hubble, pre-1950 theory of the expanding universe—there were few new data—were the Belgian astronomer Abbe' Georges LeMaître and the wonderful Russian eccentric George Gamow. In 1931 LeMaître published a paper in which he presented Friedmann's results in essentially the way that they have been presented ever since and in the way that we shall present them in this book. He also appears to have been the first scientist to draw the conclusion that if the universe is now expanding and cooling off, it must have started off from some sort of hot and perhaps singular state—what Fred Hoyle, who detested the model, later called derisively the Big Bang. LeMaître thought of the early universe as a kind of cosmic egg that had exploded and, voilá, here we are. LeMaître was not a nuclear physicist, so he was in no position to try to work out the details. Indeed, nuclear physics was in its infancy. However, George Gamow was a nuclear physicist of the very first order. Furthermore, he had intended to work with Friedmann in the early 1920s when Friedmann took a post at the University of St Petersburg. Friedmann died before Gamow could actually work with him, but Gamow seems to have been persuaded of the possibilities inherent in a theory of an expanding hot Big Bang universe from the beginning.

By the early 1940s, Gamow and his collaborators—especially Ralph Alpher and Robert Herman, who were students—had developed a serious theory of element formation in the early universe. Gamow originally hoped that if he started off with a primeval mix of elementary particles, something he came to call "Ylem" from the

Greek word *hyle* meaning either "wood" or "matter," then by invoking nuclear fusion he could account for the distribution of all the elements we actually find in the universe. Evidently, Ylem had to contain neutrons and protons—the nuclear building blocks. Since the proton is electrically charged and therefore emits radiation, Ylem will also contain energetic light quanta–gamma rays. But these can make electron-positron pairs if they are energetic enough, and so we will have electron-positron pairs as well. The neutrino, the elusive, chargeless, and perhaps massless, particle emitted in many radioactive decays, had only been conjectured by the early 1940s. It was not until 1956, in an experiment we will describe later in the book, that the neutrino was actually observed. But Gamow and his collaborators soon realized that in the early universe neutrinos would be as numerous as photons and would play an essential role. Gamow understood early in the game that the key reaction that would trigger everything else was the capture of a neutron by a proton with the formation of heavy hydrogen (deuterium) by the reaction $n + p \rightarrow d + \gamma$. Then one could build up the very tightly bound isotope of helium ^{4}He by a series of reactions like $d + d \rightarrow {}^4\mathrm{He} + \gamma$. He was under the impression for some time that one could continue this series and build up everything. The problem is that one gets stuck at nuclei with either five or eight nucleons—neutrons and protons—since these nuclei are unstable. For example, striking a helium nucleus with another one will produce ^{8}Be, an isotope of beryllium, which will promptly decay back into two helium nuclei. That aside, as heavier nuclei contain more protons, the Coulomb electrostatic repulsion between them makes it more and more difficult for them to fuse. They must have kinetic energies that are well above the Coulomb repulsive barrier energy. It turns out that at the time of nucleosynthesis, about 3 min after the Big Bang, the universe was already too cool for the successful fusion of heavier elements. The process essentially stops at ^{7}Li (lithium), which is found in trace amounts in galaxies. The heavier elements—heavier than iron—are made in the explosion of supernovae and do not have an early universe origin.

The most complete description of the work on early universe, light-element synthesis that was done by the group of people around George Gamow, was published in 1953 by Alpher, James Follin, and Herman. In reading this paper, one is constantly amazed by its prescience. At the time it was written, the amount of helium in the universe was rather imprecisely known. They quote measured value for the ratio of mass densities of helium to that of hydrogen in the universe—hydrogen being the most common element—that lie between 1:5 and 1:10. To indicate how our subject has progressed, a recent experimentally measured value of this ratio was given as 0.245 ± 0.003! Alpher, Follin, and Herman were able to fit such data as they had with their theory. When we present the theory of helium production later in the book, it will not differ in its basic essentials from the one presented in this remarkable paper, although many of the details have been refined. Even more remarkable was a paper that Alpher and Herman published in 1949. This article followed up an idea of Gamow on the fate of the electromagnetic radiation left over after the Big Bang. As we will show, until about 100,000 yr after the Big Bang, photons and electrons are kept in tight equilibrium by their rapid interactions. But by a 100,000 yr the universe has cooled sufficiently so that neutral hydrogen is formed. Electrons and protons can join to make neutral hydrogen without its being disintegrated by ambient photons. The photons are now free to expand with the universe since they no longer have significant interactions with the electrically neutral atomic hydrogen. Using an argument that we will present, Alpher and Herman showed that, at present, these photons should have a blackbody equilibrium distribution with a characteristic average temperature they estimated to be about 5 K. Again, to show the progress in the field, a recent satellite determination of this temperature gave 2.736 ± 0.01 K. As extraordinary as their prediction was, it is equally astonishing that it had no influence whatsoever on the discovery, first published in 1965, of this radiation. The saga of the accidental discovery of cosmic background radiation by Arno Penzias and Robert Wilson of Bell Telephone Laboratories has often been told. In brief, they were trying to do radio astronomy with a telescope left over from a microwave communications application. (Microwaves, with wavelengths that range from centimeters down to tenths of centimeters, are the stuff of radar, which had been extensively developed during World War II.) They discovered that their telescope was unexpectedly noisy. For a year they tried, without success, to get rid of this noise. While they were engaged in this activity, a Princeton University group headed by Robert Dicke—a few miles away—was actually constructing a radio telescope explicitly to look for cosmic background radiation. Dicke was attracted to a version of the expanding universe in which it cyclically expands and contracts. He thought that there might be background radiation left over from the last contraction. Before Dicke's group could begin their observations, they learned about Penzias and Wilson's "noise," and they realized immediately what it was. The two groups published adjoining communications in the *Astrophysical Journal* in July 1965. With these communications the modern era of cosmology began.

Before the measurement of cosmic background radiation there was little to choose between the hot Big Bang theory of Gamow and his collaborators and the Steady State theory of Hoyle and his associates. The Steady State theory is Aristotelian in that the universe is supposed to be unchanging. One could deal with expansion by introducing rather baroque notions involving the constant creation of particles, but the cosmic background radiation was too much. After it was discovered, most people lost all interest in the Steady State model. In fact, apart from these brief historical remarks, we will not discuss it at all in this book. Before the discovery of the cosmic background radiation, cosmology was not really a mainstream scientific discipline. The data were too few and far between to allow one to choose between such antithetical models as the Steady State theory and the hot Big Bang theory. Furthermore, new elementary particles were being discovered on what seemed to be an almost daily basis. They occupied the attention of most of the physicists who might have been interested in cosmology. When Alpher, Follin, and Herman wrote their paper, they did touch on the few exotic elementary particles they knew about, such as the pi and mu mesons, only to dismiss them as not playing any role in early universe light-element formation. In particular, as far as they knew, there was only one kind of neutrino, the object that is emitted in beta decay along with an electron or a positron. They did worry about whether this neutrino did, or did not, have a distinct anti particle, and they allowed for either possibility. It was not until the 1960s, however, that it became clear that there was more than one kind—"flavor"—of neutrino when a second kind was discovered. This so-called muon neutrino is emitted along with muons in decays of the pi meson. We now know that there are at least three kinds or flavors of neutrino. We also know that the elementary particles that we observe—at least the ones that interact strongly—are made up of even more-elementary particles known as quarks. There are also three known flavors of these—probably no coincidence. It turns out that the number of neutrino flavors plays an important role in determining the magnitude of the production of cosmological helium. Indeed, the experiments that have given the amount of helium in the universe with such accuracy constrain the number of neutrino flavors to be three. This is a development that occurred in the last decade. No book on modern cosmology would now be complete without a description of it and without a detailed discussion of elementary particles in general. We shall devote an entire chapter to them, and much of its sequel to their role in helium production.

The hot Big Bang theory that evolves according to Friedmann's equations and in which light-element formation takes place according to the modern version of the Gamow school's calculations, has become known as the Standard Model. It accounts for much of what is known about the early universe—from 1 s after the Big Bang to 100,000 yr. But, fortunately for the present generation of cosmologists, it is not without important loose ends. First, there is what is known as the question of the missing—or dark—matter. It turns out, and this too is a relatively recent discovery, that stars in spiral arms of galaxies do not revolve like planetary systems around a central mass. Rather, they seem to be gravitationally attracted to a sea of invisible matter in the halos of these galaxies. This mysterious matter, of an as yet unknown character, determines their orbits. It very probably constitutes most of the mass of the universe. One possibility is that it may consist of neutrinos—once again neutrinos!—if they have a small mass. This raises the prospect of massive neutrinos and their possible role in cosmology. We shall devote a whole chapter to this topic.

Another loose end in the Standard Model is what has become known as the "horizon problem." The cosmic background radiation appears to be very nearly homogeneous and isotropic. We say "very nearly" because quite recently small fluctuations, or ripples—of the order of one part in a million—have been detected in this radiation. These fluctuations may well reflect the kind of disturbances in the smooth background sea of matter and radiation in the early universe that eventually led to galaxy formation. These irregularities may be the original seeds around which the protogalaxies and clusters of galaxies clumped. But near-perfect uniformity of this background would seem to imply that all of its parts must have been at one time causally connected. Otherwise, why would all the various parts be, as we observe them today, at sensibly the same temperature? But, as we shall see, this causal connection on such a large scale is impossible if we apply the principles of general relativity and extrapolate backward using only the conventional Friedmann model of the expanding universe. To solve this dilemma, in 1980, Alan Guth introduced the notion of what has become known as "inflation." The idea is that when the universe has cooled off to about 10^{26} K then, in the blink of an eye—about 10^{-33} seconds—it expands exponentially (shades of de Sitter) by a factor of about 10^{26} and cools down to a temperature of about 1 K. In this shrunken universe, causal connectivity is possible. The universe then reinflates—the exact mechanism is still under debate—and the causal connectedness is thus explained. We shall devote a chapter to these matters.

We shall also consider the striking preponderance of matter over antimatter in those parts of the universe that we can observe. To a first approximation, there appears to be no antimatter in the present universe except that which can be produced in particle accelerators or is occasionally found in cosmic rays. We could of course simply build the absence of antimatter into our cosmology as an initial condition. But this is scientifically very unattractive. As early as 1966, Andrei Sakharov spelled out the general conditions that an explanation of this asymmetry, in the spirit of the hot Big Bang model, would have to satisfy. One feature of his scenario was that in the very early universe—at times on the order of 10^{-42} seconds—highly unstable, very massive particles existed along with their antiparticles which had the property that their decay modes violated the symmetry between particle and anti-particle. These hypothetical objects were responsible for producing the asymmetric state we now find between particles and antiparticles. After they had done their job, these conjectured "X particles" conveniently decayed away and are now nowhere to be found. While we will have the courage in this book to look back as far as 10^{-42} seconds after the Big Bang, we will not have the courage to look back farther. This would bring us to a regime in which the gravitational interactions are all powerful and very quantum mechanical. There is at present no good theory of quantum gravity, and such speculations as there are, are better suited to a textbook more advanced than this one.

As to the intended level of this book: it grew out of a one-semester course that its author gave at the Stevens Institute of Technology in the spring of 1992, 1993, and 1994. The students in the course, both juniors and seniors and first-year graduate students, were assumed to have had a sound introduction to basic physics and at least one course in what one usually calls modern physics. None of them had studied general relativity. Confronted with this limited background preparation, I had two choices: either to teach general relativity and risk never getting to the cosmological applications, or to teach cosmology and risk skimping on the relativity. What I discovered was that general relativity plays a surprisingly small role in the more interesting cosmological applications. One can, for example, give a quite decent derivation of the Friedmann equations from a purely Newtonian point of view. One can treat helium production without ever introducing a metric tensor. The one place that relativity is essential is in describing the propagation of light. To understand the horizon problem, for example, one does need a smattering of general relativity—only a smattering. Hence I have organized this book as I did my lectures: bringing general relativity in only toward the end when I needed it. I have tried to be careful about the order in which I present the topics. The first part of the book I have called the Micropaedia. It is a general overview of the subject with a good deal of emphasis on orders of magnitude. It is then followed by the Macropaedia in which the topics follow a logical progression chapter by chapter.

Teaching this course was one of the most pleasant tasks I have had as a professor, and I hope this text reflects that.

Part I

MICROPAEDIA

In this section of the book we shall present an overview of the subject. Much of the discussion will be qualitative. Most of the topics mentioned here will be treated in detail in subsequent chapters. In learning a new subject it is often useful to have a general idea of where one is going before embarking. Much of the discussion that follows will involve orders of magnitude. Hence it is very useful to have a rather complete set of units to work with. We prefer to present this tableaux as part of the text rather than as an appendix so that the reader will be motivated to become familiar with these units right from the beginning. It is also quite amusing to rewrite various familiar quantities in a variety of units. Our choice may at first sight appear arbitrary, but rest assured, all the quantities will come in handy.

1. UNITS AND ORDERS OF MAGNITUDES

We begin with a listing of fundamental constants in various units. Our base unit is the CGS system (centimeters, grams, seconds, abbreviated cm, g, sec or, when it is not confusing, s). In these units energy is measured in ergs—$\mathrm{g\,cm^2\,s^{-2}}$. However, modern cosmology involves a mixture of the macroscopic and the microscopic, which motivates us to introduce the electron volt (eV) the energy unit of the microscopic to complement the erg which is the energy unit of the macroscopic. For example, the ionization energy of hydrogen is about 13.6 eV. Thus

$$1\ \mathrm{eV} = 1.6022 \times 10^{-12}\ \mathrm{erg} = 10^{-9}\ \mathrm{GeV} = 10^{-6}\ \mathrm{MeV} \tag{1.1}$$

If we had expressed the ionization energy of hydrogen in ergs, it would have involved an absurdly small number. In these units the mass energy of a proton is 938.272 MeV.

In the same spirit we introduce the fermi (f) as the unit of microscopic length. Thus

$$1\ \mathrm{f} = 10^{-13}\ \mathrm{cm} \tag{1.2}$$

An atomic nucleus has Z protons and $A - Z$ neutrons, where A is the so-called atomic number. Typical nuclei have radii $r(A)$ given by

$$r(A) = 1.2\ \mathrm{f} \times A^{1/3} \tag{1.3}$$

Planck's constant h (which measures the scale of quantum phenomena) divided by 2π, a quantity which is called $\hbar$, is given by

$$\hbar = 1.0546 \times 10^{-27}\ \mathrm{g\,cm^2\,s^{-1}} = 6.582 \times 10^{-25}\ \mathrm{GeV \cdot s} \tag{1.4}$$

The velocity of light c is given by

$$c = 2.9979 \times 10^{10}\ \mathrm{cm\,s^{-1}} \tag{1.5}$$

In many applications the quantity $\hbar c$ enters. Thus

$$\hbar c = 3.162 \times 10^{-17}\ \mathrm{erg \cdot cm} = 197.327\ \mathrm{MeV \cdot f} \tag{1.6}$$

Modern cosmology is at the "triple point" of elementary particle physics, general relativity, and statistical mechanics. In applications of the latter, Boltzmann's constant k enters. When confusion is risked, we will call Boltzmann's constant k_b. We use K to indicate the temperature in kelvins. The Kelvin scale is the one temperature scale for which it is customary not to use a degree symbol. Hence

$$k = 8.617 \times 10^{-5}\ \mathrm{eV\,K^{-1}} = 1.381 \times 10^{-16}\ \mathrm{erg\,K^{-1}} \tag{1.7}$$

The ionization energies of atoms are, as we have noted, conveniently measured in electronvolts, so it is useful to measure temperatures in equivalent electronvolts when they are high enough so that atoms can be ionized by the ambient radiation at that temperature. More generally, it is often useful to define quantities that are "equivalent" to an energy unit, say 1 GeV. For example, the temperature equivalent to 1 GeV is 1 GeV divided by Boltzmann's constant k. Cosmologists often use the temperature and the equivalent energy interchangeably as the occasion arises. Also, by introducing suitable factors of $\hbar$ and c, we can define equivalent times, masses, and lengths. Following is a list of these equivalent quantities on scales useful for early universe cosmology.

$$1\ \mathrm{GeV}\ k^{-1} = 1.161 \times 10^{13}\ \mathrm{K} \tag{1.8a}$$

$$\frac{\hbar}{1\ \mathrm{GeV}} = 6.582 \times 10^{-25}\ \mathrm{s} \tag{1.8b}$$

$$\frac{\hbar c}{1\ \mathrm{GeV}} = 1.973 \times 10^{-14}\ \mathrm{cm} \tag{1.8c}$$

$$\frac{1\ \mathrm{GeV}}{c^2} = 1.783 \times 10^{-24}\ \mathrm{g} \tag{1.8d}$$

Often one finds cosmologists and elementary particle physicists using so-called natural units in which $\hbar = c = k = 1$. This certainly makes the formulas look more elegant, but it can be confusing when one is just learning the subject, and so we shall avoid using these units.

Early in the twentieth century, after the discovery of the quantum, Max Planck introduced a set of units for length, mass, and time that was in a certain sense universal. The foot, or the centimeter, for example, is defined in terms of earthly human lengths that one would have a great deal of difficulty explaining to an extraterrestrial. Planck wanted to introduce units based on fundamental constants whose significance one could explain to an extraterrestrial. They were not much used, but they have now come back into cosmology because of their ultimate connection with gravitation. Planck units are expressed in terms of the gravitational constant G, Planck's constant $\hbar$, and c. In earthly units the gravitational constant G is given by

$$G = 6.670 \times 10^{-8}\ \mathrm{cm^3\ g^{-1}\ s^{-2}} = 6.707 \times 10^{-39} \times \frac{\hbar c}{\mathrm{GeV^2}/c^4} \tag{1.9}$$

The first of the Planck quantities we define is the Planck mass M_P:

$$M_P = \left(\frac{\hbar c}{G}\right)^{1/2} = 2.177 \times 10^{-5}\ \mathrm{g} \tag{1.10}$$

The Planck energy $M_P c^2$ is then given by

$$M_P c^2 = 1.959 \times 10^{16}\ \mathrm{erg} = 1.221 \times 10^{19}\ \mathrm{GeV} \tag{1.11}$$

while the Planck time t_P is given by

$$t_P = \frac{\hbar}{M_P c^2} = 5.390 \times 10^{-44}\ \mathrm{s}. \tag{1.12}$$

Most physicists would agree that our knowledge of physics before the Planck time is highly speculative. The Planck length l_P is given as

$$l_P = ct_P = 1.616 \times 10^{-33}\ \text{cm} \tag{1.13}$$

Finally, the Planck temperature T_P is given by

$$T_P = \frac{M_P c^2}{k} = 1.417 \times 10^{34}\ \text{K} \tag{1.14}$$

Next we list a variety of times and ages. We shall not discuss here how they were actually measured but will return to a description of some of these measurements later in the text. We shall not give most of these ages to very high precision but rather we shall try to get a feeling for them.

$$1\ \text{yr} = 3.156 \times 10^7\ \text{s} \simeq \pi \times 10^7\ \text{s} \tag{1.15}$$

The oldest meteorite, $\sim 4.5 \times 10^9$ yr.
The oldest moon rocks, $\sim 4.5 \times 10^9$ yr.
The oldest earth rocks, $\sim 4 \times 10^9$ yr.
The age of the Solar System, $\sim 4.5 \times 10^9$ yr.
The half-life of ^{238}U, $\sim 4.5 \times 10^9$ yr.
The appearance of the first humanoids, $\sim (1\text{–}2)10^6$ yr.
The first appearance of our kind, $\sim (50\text{–}100)10^3$ yr.
The age of the universe, $t_U \sim (10\text{–}15)10^9$ yr $\sim (10\text{–}15)\pi \times 10^{16}$ s.

We now turn to a list of distances and angles.

$$1\ \text{rad} = \frac{360^\circ}{2\pi} = 57.3^\circ \tag{1.16}$$

$$1'' = 4.85 \times 10^{-6}\ \text{rad} \tag{1.17}$$

$$1\ \text{mile} = 1.609 \times 10^5\ \text{cm} \tag{1.18}$$

We introduce the symbol l_{yr} for the distance a pencil of light travels in a year. Thus

$$l_{yr} = c \times 1\ \text{yr} = 9.436 \times 10^{17}\ \text{cm} \tag{1.19}$$

$$1\ \text{km} = 1.06 \times 10^{-13}\ l_{yr} \tag{1.20}$$

If we call $R_\odot$ the mean distance from the earth to the sun, then

$$R_\odot = 1.590 \times 10^{-5}\ l_{yr} \tag{1.21}$$

It takes about 8 min for light to travel here from the sun.

Astronomers use a very peculiar distance measure called the parsec (pc). Once one gets used to it, it is a very convenient unit for measuring the huge distances that enter into cosmology. It is related to the notion of parallax. If one looks at a distant object from two different vantage points, the object will appear to shift its position against a fixed background. For example, if one looks at a star when the earth is at two different positions in its orbit around the sun, then the star will appear to shift its position with respect to a background of fixed stars. Imagine lining the sun, the earth, and the star up and then allowing the earth to move a quarter of the way around its orbit. We now have a right triangle in which $R_{\odot}$ is the base, the distance from the sun to the star is the altitude, and the distance from the earth to the star is the hypotenuse. The angle made by the altitude and the hypotenuse is the so-called angle of parallax. By measuring this angle and knowing $R_{\odot}$, one can measure, in principle, the distance to the star in question. This method is practical only when the star is not too far away. Otherwise, the angle of parallax is too small to be measurable. If we take, arbitrarily, an angle of parallax of 1″ and compute from the formula

$$\tan(1'') = \frac{R_{\odot}}{\text{parsec}} \tag{1.22}$$

the value of the parsec, we get, by definition,

$$1 \text{ pc} = 3.261 l_{\text{yr}} = 3.086 \times 10^{18} \text{ cm} = 1.918 \times 10^{13} \text{ mi} \tag{1.23}$$

We often have occasion to use the megaparsec (M pc), and we remind the reader that

$$1 \text{ M pc} = 10^6 \text{ pc} \tag{1.24}$$

Here are a couple of approximate astronomical distances measured in megaparsecs: the distance to the Andromeda galaxy, ~ 0.7 M pc; the distance to the farthest quasars (the edge of the visible universe), $\sim 3 \times 10^3$ M pc.

We might call the "size" of the visible universe, R_U, the farthest distance light could have travelled since the Big Bang. As we shall see, some models of the universe give it an infinite size, but still, in such an infinite universe, it makes sense to talk about R_U. Indeed, physically it may not make sense to talk about a greater distance since we have no way to observe it. Thus

$$R_U = ct_U \sim 10^{28} \text{ cm} \sim 10^4 \text{ M pc}, \tag{1.25}$$

which is about the distance to the farthest quasars.

We can contrast this with one of the smallest distances we will have occasion to talk about, the so-called Compton wavelength $\lambda\!\!\!^{-}_c$ associated with a particle of mass m. We will encounter this distance in several guises. It is the distance at which quantum mechanical effects become unavoidable. It is defined as

$$\lambda\!\!\!^{-}_c = \frac{\hbar}{mc} = 1.973 \times 10^{-14} \frac{\text{cm}}{mc^2/\text{GeV}} \tag{1.26}$$

In this respect the following argument is suggestive. In general relativity it can be shown that the radius of a spherically symmetric black hole of mass M is given by $2GM/c^2$. Any light trapped inside this radius cannot get out. Suppose we set $GM/c^2 = \hbar/Mc$, the Compton wavelength of a particle of mass M. We see that the above equation determines a mass and that it is the Planck mass! This seems to be telling us that at the Planck mass we ignore quantum mechanical effects at our peril. In this book we shall avoid discussing physics at the Planck mass.

2. FORCES

One of the surprising discoveries of modern physics is that there are apparently only four distinct kinds of forces that account for all the interactions we observe. Indeed, if present ideas are right, these forces may themselves be unified into a single mathematical entity. The theory we use to describe these forces is the quantum theory of fields. The most characteristic feature of this theory is that it allows for the creation and annihilation of particles. A familiar example is the electromagnetic radiation emitted by an accelerating charged particle such as an electron. This radiation is, as we know, composed of massless particles–photons. The electron acts as a source or sink for the emission or absorption of photons. In quantum field theory, the interaction of two electrons is viewed as the exchange of quanta–photons. Indeed, all particles interact by the mutual exchange of various types of quanta. Some of these quanta have mass, some have charge, some carry spin, and some have other quantum mechanical properties. The range of the interaction between two particles is related to the mass of the quanta exchanged by them. This is a consequence of the uncertainty principle. When a quantum of mass m is emitted, energy conservation is violated by an amount mc^2. This is allowed, provided it happens for a sufficiently short interval; i.e.,

$$\Delta E \Delta t \sim \hbar \tag{1.27}$$

Thus

$$\Delta t \sim \frac{\hbar}{mc^2} \tag{1.28}$$

These quanta move essentially at the speed of light, so that in the interval Δt they move a distance $\hbar/mc$—a Compton wavelength. This is the range of the force in question. The strength of the forces is characterized by what is known as a coupling constant. On page 9 is a diagram—a so-called Feynman graph—that shows a particle interacting by the exchange of quanta. At the vertex where the particle emission or absorption takes place, we have indicated the coupling constant in conventional units. Defined in these units, the square of the coupling constant divided by the product $\hbar c$ is a dimensionless number.

It is customary, as we have shown in Fig. 1, to rationalize the value of the coupling constant by dividing its square by 4π. The strongest of the four forces—the strong force that holds the nucleus together—is characterized by a coupling constant $g^2/4\pi\hbar c \sim 1$. For nuclear phenomena at low energies, a theory in which the nucleons—neutrons and protons—exchange a spinless meson called the pi meson with a mass energy of $m_\pi c^2 = 139.568$ MeV gives a satisfactory account. Using Eq. 1.26, we see that this mass corresponds to a range of about 1.4 f which, according to Eq. 1.3, is about the size of a typical nucleus. For high-energy nuclear phenomena the so-called quark model is more suitable. Here, each nucleon is made of still more fundamental particles called quarks and these interact by the exchange of mesons called "gluons". The coupling constant given above is the quark-gluon coupling constant. We shall have occasion to return to the quark model in a later chapter.

After the strong force, the next strongest force is the electromagnetic force. It is produced when electrically charged particles exchange photons. This process is characterized by a dimensionless coupling constant $\alpha = e^2/4\pi\hbar c = 1/137.036$. Here e is the charge of the electron. We have chosen units in which the permittivity of free space ϵ_0 has been set equal to 1. We can define the so-called classical radius of the electron in terms of α and the mass energy of the electron, $mc^2 = 0.51$ MeV. By definition, the classical radius of the electron, r_0, is given by the equation that sets the Coulomb and the rest energies of the electron equal; i.e.,

$$\frac{e^2}{4\pi r_0} = mc^2 \tag{1.29}$$

or, solving,

$$r_0 = \alpha \lambda\!\!\!^{-}_e = 2.818 \text{ f} \tag{1.30}$$

Next on the list of interactions is the weak interaction. Weak interactions are responsible for the decay of many elementary particles. These interactions will play a profound role in our understanding of cosmology. A prototypical weak interaction is the decay of the neutron. The neutron has a rest energy of $m_n c^2 = 939.566$ MeV, while the proton, as we have mentioned, has a rest energy of $m_p c^2 = 938.272$ MeV. Thus the neutron-proton rest energy difference is 1.293 MeV. This means that it is energetically possible for a neutron to decay into a proton plus an electron of rest energy $m_e c^2 = 0.511$ MeV. If neutron decay were really a two-body process, then the energy-momentum kinetics would predict that the electron would emerge with a single energy determined by the rest masses. But this is not what is observed. The electron emerges with a spectrum of energies. The simplest explanation of this is that the electron is emitted along with another particle. Whatever that particle is, it must be electrically neutral since the electron and proton charges balance and the neutron is electrically neutral. This neutral particle is called the neutrino. We shall have occasion to examine its properties in detail later. For present purposes we shall simply call it ν and write the decay of the neutron as

$$n \rightarrow p + e^- + \nu$$

We have written a minus sign over the e because the electron has an antiparticle called the positron with the same mass but opposite charge. We shall designate the positron by the symbol e^+. The neutrino also has an antiparticle which we designate by $\bar{\nu}$. Strictly speaking, the particle that is emitted with the electron is an antineutrino. This subtlety we shall also take up later.

The first theory of weak interactions was developed in the early 1930s by Enrico Fermi. Within the limits of its applicability it was a very successful theory, so we still retain some of its conventions even though we now understand its limitations. The essence of Fermi's theory was that it involved a contact interaction among the particles as shown in Fig. 2.

If this were a correct description of the process, it would, in our previous language, involve the exchange of a meson of infinite mass—something that is unphysical. One of the important discoveries of the last 20 yr is that there are mesons of finite—albeit large—masses that transmit this interaction. We shall return to the properties of these mesons shortly. Here we shall use the conventions of the Fermi theory which applies so long as the energies involved are small compared to the masses of the intermediate weakly interacting mesons. One of the most important quantities to measure in the decay of the neutron is its lifetime. We remind the reader that if the number of neutrons in a sample at time t is $N(t)$, then $N(t)$ obeys the differential equation

$$\frac{\mathrm{d}N(t)}{\mathrm{d}t} = -\frac{1}{\tau} \times N(t) \tag{1.31}$$

where τ is the neutron's lifetime. We also remind the reader that τ is related to the half-life (the time it takes for half the sample to decay) by the equation

$$t_{1/2} = 0.693\tau \tag{1.32}$$

We can derive Eq. 1.32 from the solution to Eq. 1.31, namely,

$$N(t) = e^{-t/\tau} N(0) \tag{1.33}$$

In principle, knowing this solution we can determine τ from experiment, although in practice it may be difficult. For the neutron decay the best value of τ is now given as

$$\tau = 888.6 \pm 3.5 \text{ s} \tag{1.34}$$

The Fermi theory gives a relation between τ and the strength of the Fermi coupling constant G_F. In order not to be obscure we shall write down this relation in the

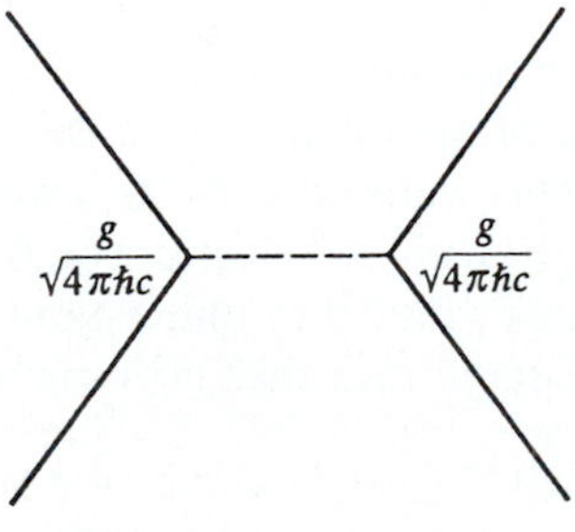

Figure 1 Two particles exchanging a field quantum.

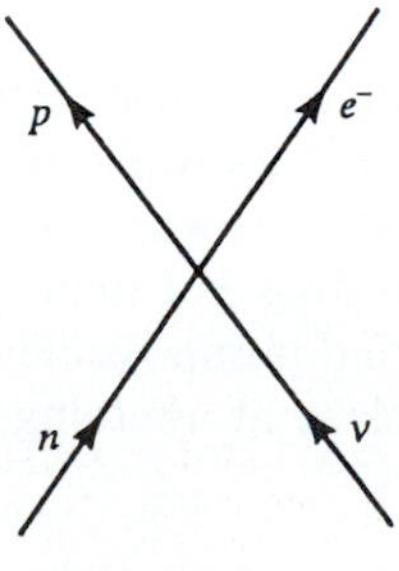

Figure 2 The four-fermion contact interaction that produces $n \rightarrow p + e^- + \nu$.

simplest approximation. It would take us too far afield to derive it, and we refer the reader to the standard texts on the quantum theory of fields. Thus

$$\frac{1}{\tau} = \frac{G_F^2}{60\pi^3} \times \frac{c^4}{\hbar^7} \times m_e^5 \left(\frac{m_n - m_p}{m_e} \right)^5 \tag{1.35}$$

The reader will note the strong dependence of the lifetime on the neutron–proton mass difference or, more generally, on the energy release in the decay. That is one reason why there is such a wide range of observed lifetimes for particle decay. The energy releases can be very different. Using the more sophisticated versions of Eq. 1.35 and knowing τ, we can find G_F. Because we have used a contact interaction, G_F will be expressed in terms of strange-looking units. Later, when we discuss the intermediate meson theory of weak interactions, we will introduce a weak coupling constant that is dimensionless—like α. In the strange units,

$$G_F = 8.923 \times 10^{-47}\ \text{GeV} \cdot \text{cm}^3 = 1.430 \times 10^{-49}\ \text{erg} \cdot \text{cm}^3 \tag{1.36}$$

It is useful to rewrite Eq. 1.36 by introducing $\hbar$ and c. Thus

$$\frac{G_F\,\text{GeV}^2}{(\hbar c)^3} = 1.162 \times 10^{-5} \tag{1.37}$$

Not long after Fermi introduced his theory, the Japanese physicist Hidekei Yukawa, who was a pioneer in the pi meson theory of strong interactions, suggested that the pi meson might also transmit weak interactions. This turned out not to work. The reason why it did not work teaches us something about the particles and their couplings. A particle like the neutron has strong, weak, and electromagnetic interactions. All massive particles have gravitational interactions. However, the electron has no strong interactions, but it does have weak and electromagnetic interactions. The neutrino, on the other hand, has weak interactions and no strong or electromagnetic ones. (The electromagnetic interactions of neutrinos are a bit tricky. It is more correct to say that they have no primary electromagnetic interactions.) A process like electron neutrino scattering

$$e^- + \nu \rightarrow e^- + \nu$$

then takes place primarily through weak interactions, with tiny electromagnetic corrections. In Yukawa's picture, the pi meson would be the intermediate state in the process $e^- + \nu \rightarrow \pi^- \rightarrow e^- + \nu$. This would involve two weak interactions since the pi meson does not interact strongly with the electron or the neutrino. But one would then find it impossible in Yukawa's model to understand how this interaction could take place at anything like the same rate that governs a process like

$$\nu + n \rightarrow e^- + p$$

the inelastic scattering of a neutrino from a neutron, where the pi meson can interact strongly with nucleons; i.e., in this theory this reaction would involve one strong and one weak coupling. In actual fact, the rates of the two processes are comparable. There is a kind of universal strength for these weak interactions. Once this universality was appreciated, it became clear that a new kind of meson—a so-called weak meson—was needed to transmit weak interactions. It had to be very massive because Fermi's four-point coupling theory works so well at low energies. It also had to be electrically charged to produce the neutrino–neutron inelastic process described above. Furthermore, it had to have no strong interactions so as to avoid the problem the pi meson presented. It was given the name W (weak) meson, and it transmits weak interactions by processes like the one depicted in Fig. 3. We have introduced the weak coupling, dimensionless coupling constant g_W at the appropriate vertices.

Theorists also conjectured that there should be a neutral counterpart to the $W^{\pm}$. It was called the Z^0, and it plays a role in processes like $\nu + p \rightarrow \nu + p$ via diagrams like Fig. 4.

In the 1980s both of these types of mesons were actually observed in experiments done at high-energy accelerators. It turns out that the mass energies are, respectively,

$$m_W c^2 = 80.6 \pm 0.4 \text{ GeV}$$

and

$$m_{Z^0} c^2 = 91.161 \pm 0.031 \text{ GeV}$$

If we work out the theory of weak interactions involving these mesons and then demand that it give the same results at low energy as the Fermi theory, we can find the value of g_W. The relation between g_W and G_F, which is far from obvious, turns out to be

$$\frac{G_F}{(\hbar c)^3 2^{1/2}} = \frac{g_W^2}{4\pi\hbar c} \times \frac{1}{8 m_W^2 c^4} \tag{1.38}$$

Using Eq. 1.37 in Eq. 1.38, we find that $g_W^2/4\pi\hbar c \simeq 10^{-2}$. This is of the same order of magnitude as α, the fine structure constant, which suggests that there might be a connection between weak interactions and electromagnetism. Indeed, there is such a connection. Both theories are subtexts of a more general unified electro-weak theory which has been developed in the last few decades. It is discussed in all standard texts on elementary particles.

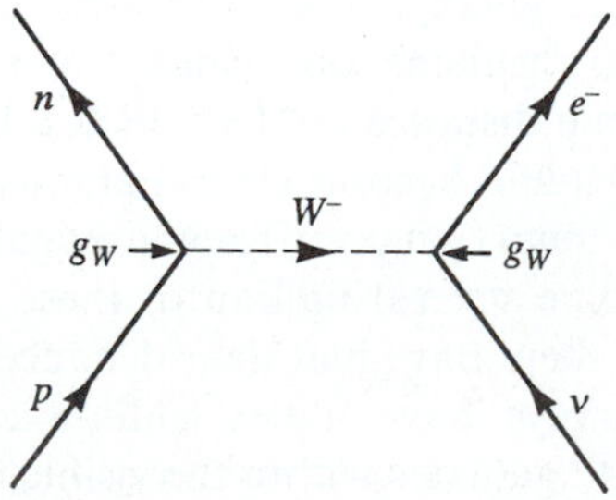

Figure 3 A Feynman diagram for the process $\nu + n \rightarrow e^- + p$ via the exchange of a W^- meson.

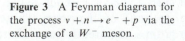

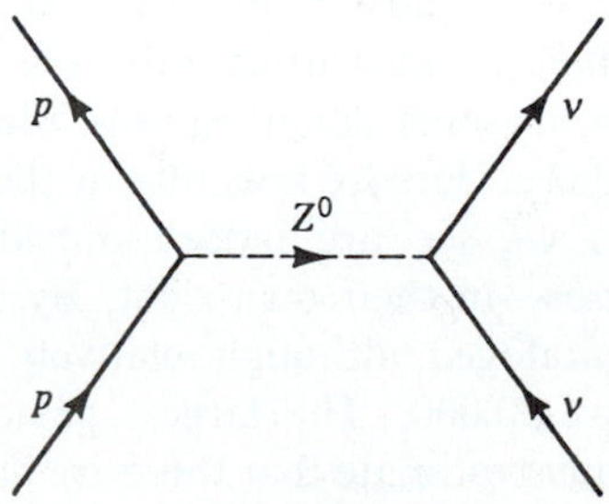

Figure 4 A Z^0 exchange in neutrino proton scattering.

The final force that we know about is gravitation. From a particle physics point of view it is the weakest force we know. Consider, for example, two protons at a distance r from each other, where r is large enough so that quantum mechanical effects can be ignored. Then, using Eq. 1.9 and recalling that the mass energy of the proton is about 1 GeV, while $e^2/4\pi\hbar c \simeq 1/137$, we have

$$\frac{GM^2/r^2}{e^2/4\pi r^2} \simeq 10^{-36}$$

The reason why gravitation appears to be so effective is that the matter we see around is by and large electrically neutral so that the gravitational force dominates. At present there is no truly successful quantum theory of gravity. This theory, when it is produced, will be needed to describe the physics around the Planck time and earlier. We will not enter into these highly speculative matters in this introductory text.

3. WHAT IS THERE?

At the very least, cosmology must account for what we observe the cosmos to be made up of. Of course, there are limits. We do not expect cosmology to account for the existence of tigers. However, we certainly expect it to account for the fact that there are about 400 microwave photons per cubic centimeter, on average, in every cubic centimeter in the universe. We also expect it to account for the fact that about a quarter of the mass of the universe is in helium. In addition, we expect it to account for the gross features of the formation of galaxies, although we do not expect it to account for the number of moons of Jupiter. In this section we shall describe a few of the things that cosmology should account for in preparation for the detailed treatment to follow.

3.1 GALAXIES

Until early in this century, it was generally believed by astronomers that all the stars we see were part of our own Milky Way galaxy. There were no good methods available for measuring distances to stars outside our galaxy. We shall shortly discuss, in some detail, the so-called cosmic distance ladder which takes us outside our galaxy. Here we note that by the 1920s it had become clear that most of the stellar objects we see are indeed outside our own galaxy. They are galaxies—island universes—in their own right. By now some several million of these galaxies have been cataloged, although relatively few of them have had their distances measured—perhaps 30,000. The largest galactic surveys have a few million examples. Astronomers estimate that there are about 10^{11} galaxies within the visible universe, each with about 10^{11} stars. The typical size of such a galaxy is about 20 kpc. Only about 10 to 20 percent of these galaxies appear to stand on their own. Those that do are called "field galaxies." The rest stand in groups, apparently attracted to each other by their mutual gravitation. For example, it appears as if the Andromeda and the Milky Way galaxies are attracted to each other and are moving toward each other with a relative speed of about 300 km s^{-1}. But there is a hierarchy of nearby galaxies, about 30 of them, which appear to be correlated. They are known as the "local group." If a system of galaxies has more than 100 members it is known as a "galactic cluster." The ones nearest to the Milky Way are the so-called Virgo and Coma clusters. The Virgo cluster, which is about 15 Mpc away from us, contains in its 3-Mpc diameter several thousand galaxies. About 10 percent of the galaxies occur in clusters. A striking feature of both galaxies and galaxy clusters is that their distribution in space appears, on the average, to be nearly uniform. Astronomers tell us that if we take any cube in the observable universe with a side of about 50 Mpc, then its properties—the matter density and the rest—will be sensibly the same as for any other cube of the same size. Recall that the size of the visible universe is about 10^4 Mpc, which gives some notion of how good this uniformity assumption is supposed to be. For many purposes, including cosmology, a good first approximation to the distribution of stellar matter in the universe is to consider it to have a uniform density. To estimate this density we have to know both the mass and volume of the universe. As we have mentioned, in some models the volume of the universe might actually be infinite. However, the volume of the visible universe, V_U, is always finite. We can write

$$V_U = \frac{4\pi}{3} R_U^3 \tag{1.39}$$

where R_U is given by Eq. 1.25, that is,

$$V_U \sim 10^{84}\ \text{cm}^3$$

Obviously we cannot weigh a galaxy or a galaxy cluster by putting it on a scale. We will, however, give the flavor of one of the methods used indirectly to measure the mass of a galaxy cluster. In these clusters a single galaxy can sometimes be treated as an isolated mass that rotates around the nucleus of the cluster like a planet revolving around the sun. The linear speed at which this rotation takes place can be measured using, for example, Doppler shifts. Typical galactic motions in a cluster are on the order of 1000 km s^{-1}. We can also measure the average distance the galaxy in question is from the central mass—some tens of megaparsecs. Now we can apply, at least in an average sense, the equations for uniform circular motion under the influence of gravity,

$$ma = \frac{mMG}{r^2} = \frac{mv^2}{r} \tag{1.40}$$

where G is Newton's constant, m is the mass of the galaxy, and M is the mass of the cluster. We can now solve for M. The useful unit in which to express all these galactic masses is the mass of the sun, $M_\odot$, where

$$M_\odot = 1.989 \times 10^{33}\ \text{g} \tag{1.41}$$

A typical galaxy cluster has a mass of about $10^{15}\, M_{\odot}$. If we add all these masses up and divide by V_U, we can estimate the density of visible matter in the visible universe, a quantity we will call ρ_m. These estimates yield

$$\rho_m \simeq 5 \times 10^{-31} \frac{\text{g}}{\text{cm}^3} \tag{1.42}$$

This corresponds to an energy density ρ_E given by

$$\rho_E = \rho_m c^2 \simeq 4.5 \times 10^{-10} \frac{\text{erg}}{\text{cm}^3} \tag{1.43}$$

Most of this matter is in the form of hydrogen. Since we know the mass of the proton, m_p, in grams,

$$m_p = 1.673 \times 10^{-24}\ \text{g}$$

we can re-express ρ_m in terms of proton masses. Thus

$$\rho_m \simeq 3 \times 10^{-7} \frac{m_p}{\text{cm}^3} \tag{1.44}$$

If we multiply ρ_m by V_U, we can get a crude estimate of the number of protons in the visible universe, N_p:

$$N_p \simeq 10^{77}$$

3.2 MICROWAVE PHOTONS

We assume the reader has some familiarity with the theory of blackbody radiation. In particular, one should recall that when radiation is distributed in a blackbody spectrum, it means that the radiation, and the charged matter it scatters from, have achieved thermodynamic equilibrium at some temperature T. We shall devote an entire chapter to the subject of equilibrium, and we shall demonstrate why the cosmic background radiation is in an equilibrium distribution. Here we will take it as an empirical fact that such a distribution exists. It turns out to have a temperature of $T = 2.736 \pm 0.01$ K.

This astonishingly precise number was determined in 1992 by using the Cosmic Background Explorer (COBE) satellite which carries radiometers well above the earth's atmosphere. The experimental curve fits a theoretical blackbody distribution to within a fraction of a percent. For such a distribution, all its properties are uniquely determined once the temperature is known. One such property is the wavelength of the radiation corresponding to the maximum in the wavelength distribution, λ_{max}. This is, in some sense, the most likely wavelength at which the radiation will be found, although there will be radiation at all wavelengths. This maximum wavelength is given by the so-called Wien displacement law which states that

$$\lambda_{\text{max}}\, T = 0.290\ \text{K} \cdot \text{cm} \tag{1.45}$$

Thus λ_{max} for the cosmic blackbody radiation is on the order of $\frac{1}{10}$ cm, putting it in the microwave regime and thereby making it accessible to radio telescopes.

The temperature also determines the average number of photons per cubic centimeter, n_γ. The formula for this, counting both states of photon polarization is

$$n_\gamma = \frac{2.4}{\pi^2}\left(\frac{kT}{\hbar c}\right)^3 \tag{1.46}$$

Putting in the numbers, we find that for 2.74 K,

$$n_\gamma = \frac{417}{\text{cm}^3}$$

We are swimming in microwave photons. We can also compute the energy density of these photons, ρ_γ:

$$\rho_\gamma = \frac{\pi^2}{15}\frac{(kT)^4}{(\hbar c)^3} \simeq 2.7kT \times n_\gamma \tag{1.47}$$

If we put in 2.74 K, we find that

$$\rho_\gamma = 4.77 \times 10^{-13}\ \text{erg cm}^{-3}$$

In other words, the energy content of these microwave photons is about 1000 times less than the energy content of the visible matter. It is amusing to estimate the total number of these photons in the visible universe, N_γ. We find

$$N_\gamma = n_\gamma V_U \simeq 10^{86}$$

Roughly speaking, the total photon entropy S_γ in the visible universe is given by N_γ multiplied by Boltzmann's constant, so that

$$S_\gamma \simeq 10^{70}\ \text{erg K}^{-1}$$

We may define σ, a dimensionless number, so that σ multiplied by Boltzmann's constant k (σk) is the photon entropy per baryon (per neutron or proton) in the universe. In the present universe, we see from our previous discussion that σ is a very large number, on the order of 10^7. We will later show that during most of the life of the universe, this number has been constant and very large. When the specific entropy is large, there is a great deal of heat. This is why we refer to what has become the standard cosmological scenario as the hot Big Bang model.

3.3 NEUTRINOS

If present cosmological ideas are correct, there are almost as many neutrinos left over from the Big Bang as there are photons. Later on we shall explain why we believe this to be true. Since these neutrinos have only weak interactions, they have so far eluded detection. If they have even a small mass, compared to, say, the mass of the electron, they could constitute the dominant contribution to the energy density of the present universe. We will leave the discussion of these intriguing matters for the sequel.

3.4 HELIUM

One of the real triumphs of modern cosmology has been the quantitative account it gives of the amount of helium found in the present universe. This is a complicated and fascinating story, and we shall give some of the preliminaries here, reserving the rest for later in the text when we have more of the machinery assembled. We define M_{He} and M_{H} to be the total amount of mass of helium and hydrogen, respectively, in the universe. It is customary to define the quantity Y as

$$Y = \frac{M_{\text{He}}}{M_{\text{H}} + M_{\text{He}}} = \frac{1}{1 + M_{\text{H}}/M_{\text{He}}} \tag{1.48}$$

Examining the helium content of both galactic and extragalactic regions—we shall say more about this when we treat helium production in detail—astronomers have concluded that

$$Y \simeq 0.23 - 0.24$$

Before giving the argument that all this helium cannot be produced in stars, we will first convert this result for Y into a result for the relative number of helium and proton nuclei in the universe. The stable isotope of helium is ^{4}He, which consists of two neutrons and two protons. We can write its mass energy as

$$m_{\text{He}}c^2 = 2m_n c^2 + 2m_p c^2 - B \tag{1.49}$$

where we have introduced the binding energy B, which we take as a positive number. In what follows, we will ignore the neutron–proton mass difference, as compared to the masses of the particles themselves, so that we can write

$$m_{\text{He}}c^2 \simeq 4m_p c^2 - B \tag{1.50}$$

Since the binding energy of ^{4}He is only 28.28 MeV, compared to the gigaelectronvolt nucleon mass energies, we will ignore it in Eq. 1.50. Thus we can write

$$Y \simeq \frac{4N_{\text{He}}}{N_{\text{H}} + 4N_{\text{He}}} = \frac{1}{(N_{\text{H}}/4N_{\text{He}}) + 1} \tag{1.51}$$

If we solve this for $N_{\text{He}}/N_{\text{H}}$, we find $N_{\text{He}}/N_{\text{H}} \simeq 0.075$.

We now ask whether this much helium, either by mass or number, can be accounted for by the processes that manufacture helium in stars. To come to grips with this matter we must first ask how helium is manufactured in stars. There are basically two nuclear fusion mechanisms—the proton-proton cycle and the carbon cycle—which in effect turn four protons into one ^{4}He nucleus. The proton–proton cycle, which is operative for cooler stars like our sun, begins with the weak reaction

$$p + p \to d + e^+ + \nu$$

where d is the nucleus of heavy hydrogen (the deuteron) which consists of a neutron and a proton bound together with a binding energy of 2.23 MeV. This reaction is followed by

$$d + p \rightarrow {}^3\text{He} + \gamma \ (5.51 \text{ MeV})$$

The energy in parentheses is the energy released to the helium nucleus and the gamma ray because of the difference in rest masses on the two sides of the reaction equation. This mass difference is what supplies the energy released in the reaction. But this reaction is followed by

$${}^3\text{He} + {}^3\text{He} \rightarrow {}^4\text{He} + 2p + \gamma \ (12.98 \text{ MeV})$$

Hence four protons have been converted into a ^{4}He nucleus and two protons. In this process a total of about 18.5 MeV has been released.

The carbon cycle is somewhat more elaborate. We list the reaction sequence and the principal energy releases below, using the standard abbreviations C for carbon, N for nitrogen, and O for oxygen. Thus

$${}^{12}\text{C} + p \rightarrow {}^{13}\text{N} + \gamma \ (1.93 \text{ MeV})$$

$${}^{13}\text{N} \rightarrow {}^{13}\text{C} + e^+ + \nu$$

$$p + {}^{13}\text{C} \rightarrow {}^{14}\text{N} + \gamma \ (7.6 \text{ MeV})$$

$$p + {}^{14}\text{N} \rightarrow {}^{15}\text{O} + \gamma \ (7.39 \text{ MeV})$$

$${}^{15}\text{O} \rightarrow {}^{15}\text{N} + e^+ + \nu$$

$${}^{15}\text{N} + p \rightarrow {}^{12}\text{C} + {}^4\text{He}$$

The carbon in this reaction has acted as a catalyst. We start with carbon, and we end with carbon. On the average, these two stellar helium manufacturing processes release about 18 MeV of energy, largely in photons. We may ask, If all the helium we see were made this way, what would be the energy density of the emitted photons? We may then compare this number to the actual observed photon radiation density from starlight. We emphasize that these photons are not the cosmic background photons that were made not too long after the Big Bang and have had eons to cool down. Starlight photons were made in the recent past and are still being made. We may write the equation

$$\frac{\text{Energy liberated in } \gamma\text{'s}}{\text{centimeter}^3} = \text{average energy liberated per helium formation} \times \frac{\text{number of helium nuclei}}{\text{centimeter}^3} = \bar{E} n_{\text{He}} \quad (1.52)$$

Here $\bar{E}$ stands for the average energy released, and n_{He} is the density of helium nuclei. From Eq. 1.44 we see that the nucleon density in the universe n_{nuc} is given by

$$n_{\text{nuc}} = 3 \times 10^{-7} \text{ cm}^{-3} \quad (1.53)$$

But these nucleons, apart from trace amounts, are either in hydrogen or helium. Thus, since there are four nucleons in each ^{4}He nucleus,

$$n_{\text{nuc}} = n_{\text{nuc in He}} + n_p = 4n_{\text{He}} + n_p \tag{1.54}$$

However, following the discussion of Eq. 1.51, we know that the ratio of the number of helium nuclei to the number of free protons is about 0.075. Thus we can solve for the helium density to find

$$n_{\text{He}} \simeq 1.7 \times 10^{-8}\ \text{cm}^{-3}$$

We can put this result into Eq. 1.52, taking 18 MeV as the average energy released, to learn that the energy density per centimeter cubed that would be emitted if all the observed helium were recently manufactured in stars would be about 5.0×10^{-13} erg cm^{-3}. But the *total* starlight energy observed is only about 0.09×10^{-13} erg cm^{-3}. This is two orders of magnitude less than what should be observed if helium production were a result of recent nuclear processes in stars. This may not prove that the helium we see has a cosmological origin, but it is suggestive. Finally, we note that there are trace amounts of other light nuclei whose origin is probably cosmological. In particular, about 1 in 10^5 nuclei is a deuteron—a heavy hydrogen nucleus. These cannot be of recent stellar origin since any deuterium made in stars is rapidly consumed by the reaction

$$d + d \rightarrow {}^4\text{He} + \gamma$$

The conscientious student should attempt to figure out why the same kind of argument does not apply to helium produced in stars.

3.5 WHAT YOU DON'T SEE YOU GET ANYWAY

One of the more important discoveries of the last few decades is that some—perhaps most—of the matter of the universe is invisible—"dark matter." This characterization presumably reflects the present level of our technology. For example, one of the candidates for this dark matter are neutrinos—possibly massive. Cosmological neutrinos are "invisible" only because no one has produced a suitable apparatus for detecting them. When this is done, they will become as visible as anything else. The same remark applies to the other candidates for the dark matter such as weakly interacting massive particles (WIMPs), which might be objects other than neutrinos. In this section we want to explain a few of the reasons why we believe that this dark matter, whatever it is, exists. To this end, we must say a few words about galaxies. This is a huge subject in its own right and could well take up an entire course. We shall only scratch the surface here.

When one looks at photographs of galaxies, they seem to come in a bewildering variety of shapes. However, these shapes have been classified generically in the Hubble classification, named after the American astronomer Edwin Hubble whose observations in the 1920s and 1930s opened up much of the modern era in cosmology. The Hubble classification is shown in Fig. 5.

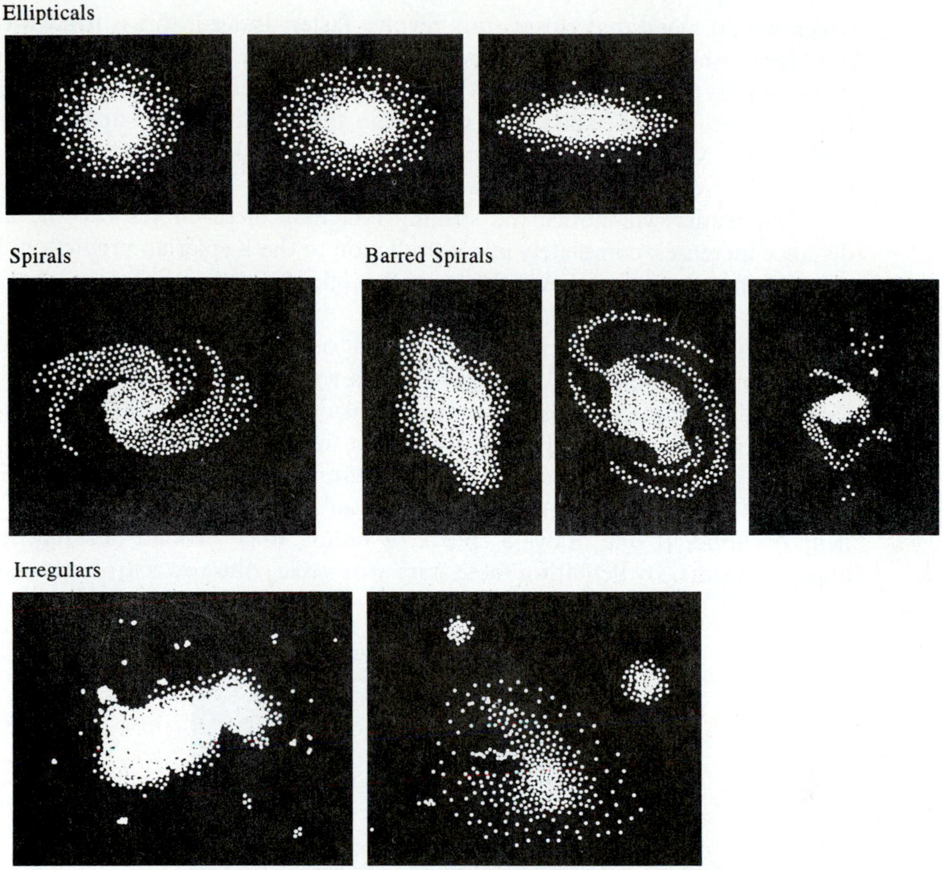

Figure 5 Ellipticals Spirals Barred spirals Irregulars

Our own Milky Way galaxy is a spiral which looks edge-on as seen in Fig. 6.

Galaxies rotate, presumably reflecting something about how they were formed. The stars in the spiral arms revolve around the galactic nucleus at speeds of a few hundred kilometers per second. There are two sorts of circular motion one could imagine: a rigid rotation, as if these stars were attached to a gigantic wheel, or a Keplerian revolution like that of the planets around the sun (Fig. 7). In the latter case the stars would revolve independently of each other but under the influence of a common gravitational field.

The latter case is governed by the Keplerian laws (Eq. 1.40). Here the velocity of revolution is related to the distance from the galactic nucleus by the equation

$$v^2 = \frac{MG}{r} \tag{1.55}$$

which would mean that closer stars revolve faster. Indeed, this is what was claimed in older cosmology books. However, in the last 20 yr or so, it was discovered that this is simply not true. These new measurements determine the revolutions of stars found in the halos of spiral galaxies where there is little light apart from the odd visible star. Most of the matter in these halos is dark. In Fig. 8 we plot a typical rotation curve for stars in such halos.

The reader will notice the striking fact that these curves become flat as the distance increases, completely in contradiction to the Keplerian prediction. For some galaxies the curve turns upward for small r—the Keplerian behavior—but is flat for large r as shown in Fig. 9.

The most obvious explanation of this is that the picture of stars under the gravitational attraction of a central mass is wrong. In fact, the empirical results is that $M \sim r$. In other words, the invisible galactic matter extends well into the halo. A better picture of such a galaxy is shown in Fig. 10. One of the great unsolved mysteries of modern cosmology is the nature of this dark matter.

Another illustration of the presence of dark matter comes from our own neighborhood. If one draws a sphere of radius 10 pc around our sun, it contains about 300 stars. By definition these stars are visible. But now astronomers divide this

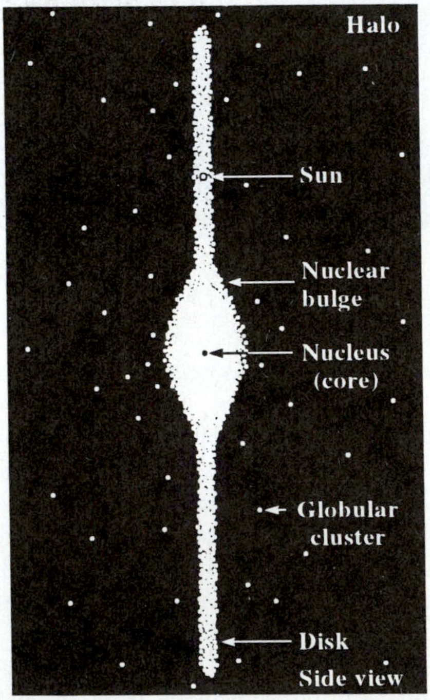

Figure 6 The Milky Way galaxy edge-on. Most of its stars lie in a flat disk. The sun is located at about 8.5 kpc from the center and travels in a circular orbit around the center with a speed of about 220 km s^{-1}.

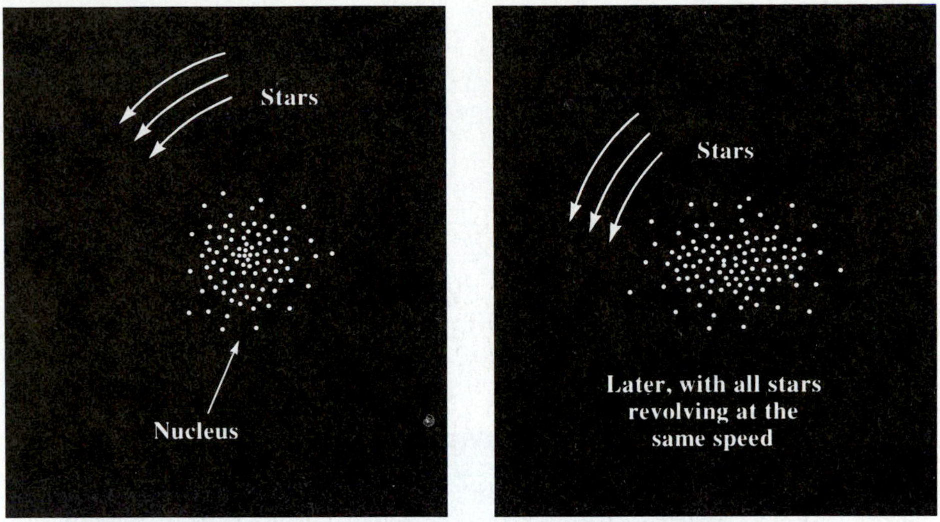

(a) (b)

Figure 7a Stars rigidly revolving.

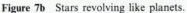

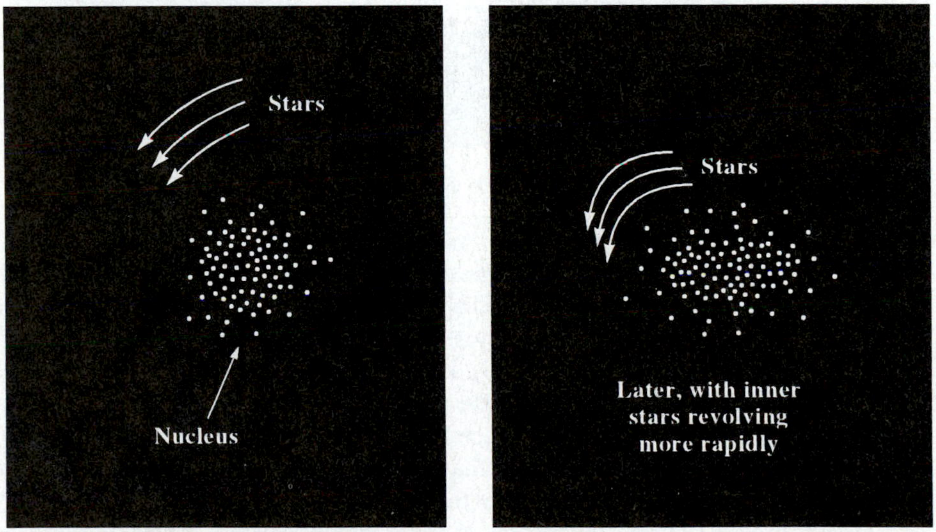

(a) (b)

Figure 7b Stars revolving like planets.

sphere into two pieces. There are about 61 stars within 5 pc and about 239 stars between 5 and 10 pc. Working out the corresponding densities, one sees that in the larger volume the density has dropped by a factor of 2. It seems very unreasonable that there should be such a sharp falloff in density over such a small region of space. The natural assumption is that material not visible to us provides the missing mass. One possibility involves what astronomers call brown dwarfs, stars that shine so

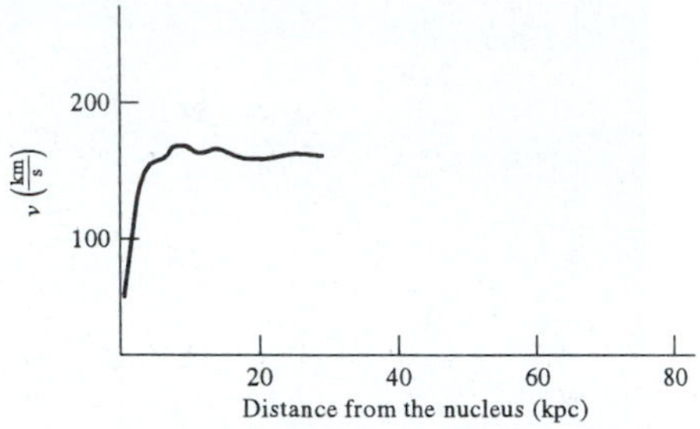

Figure 8 The galactic rotation curve for stars in the halo of a spiral galaxy.

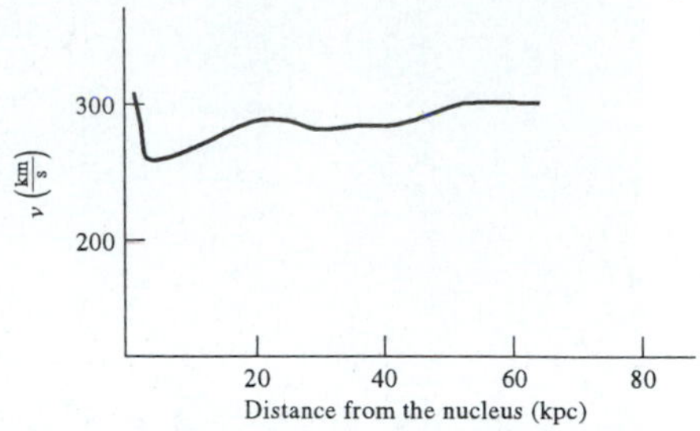

Figure 9 A rotation curve showing Keplerian behavior for small *r*.

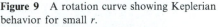

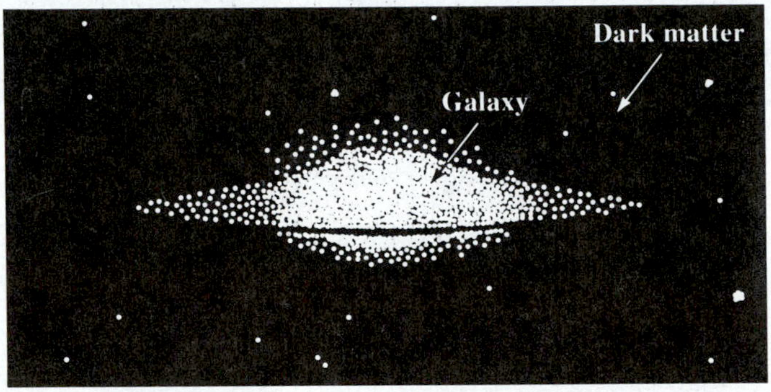

Figure 10 A spiral galaxy with its halo of dark matter.

faintly we can't observe them. Astronomers also introduce what they call the mass-to-light ratio Υ. This is a measure of the ratio of the mass to the luminosity of an object or a region in space. To see how it works consider the sun. Its mass $M_\odot$ is 1.989×10^{33} g, while its luminosity $L_\odot$ is 3.90×10^{33} erg s^{-1}. In units of gram seconds per erg, $\Upsilon_\odot = M_\odot / L_\odot$ is about 0.5. In the neighborhood of the sun the mass-to-light ratio is thought to be about $1.5\Upsilon_\odot$. This ratio has been estimated for various distant astronomical objects, galactic clusters being a striking example. Here one takes advantage of the fact that galaxies in a cluster move with a distribution of radial velocities determined by the mass they experience. This gives a dynamical method of measuring the masses of these galaxy clusters. Their luminosities can be measured more directly. In the Coma cluster, for example, this method produces a mass-to-light ratio of about $300\Upsilon_\odot$! Hence the stars that we see in the Coma cluster account for only a few percent of its mass. The rest must be dark matter whose nature is as yet mysterious.

This completes our brief overview. In the next section of the book we examine these topics and others in detail.

Part II

MACROPAEDIA

Having given the reader a feeling for a few of the broad ideas in cosmology, we want, in this part of the book, to examine the important topics in detail. We present these topics in such a way that the information in the precedent discussions will be relevant to the subsequent ones. We begin with the basic question of why we believe that the universe is expanding.

1

The Expanding Universe—Phenomenology

We start by reminding the reader of a few facts from the special theory of relativity. This theory deals with the relationships of quantities like space, time, and mass in two coordinate systems S and S′ that are moving with respect to each other with a uniform velocity v (Fig. 1).

If S measures the length of a ruler at rest with respect to his system to be L, then S′ will claim that the same ruler has a length L' given by

$$L' = L\sqrt{1 - \frac{v^2}{c^2}} \tag{1.1}$$

and vice versa. Conversely, if S claims that a clock at rest in his system beats with a frequency ν, then S′ will claim that this clock beats with a frequency ν' given by

$$\nu' = \nu\sqrt{1 - \frac{v^2}{c^2}} \tag{1.2}$$

These two facts enable us to derive the formula for the relativistic Doppler shift. We do this for the case in which the observer is at rest and the source in motion. We

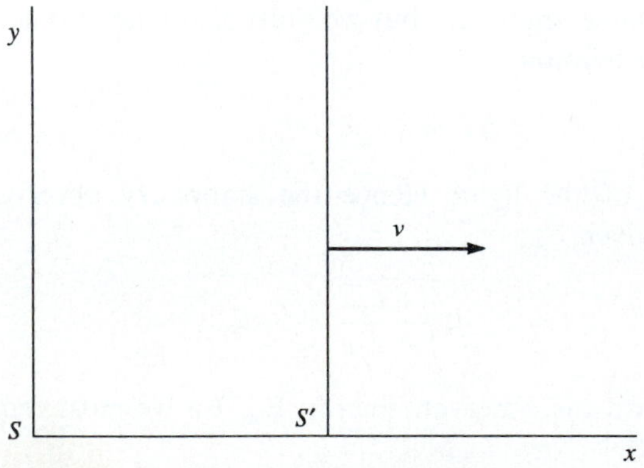

Figure 1 Two systems S and S′ moving with a velocity v with respect to each other along the x axis.

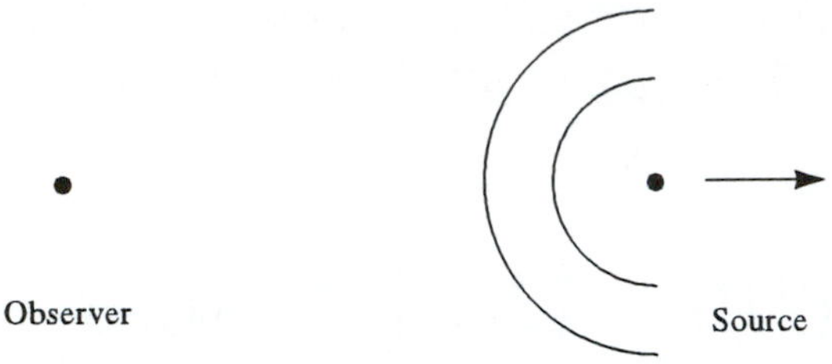

Figure 2 A source moving away from an observer while emitting spherical light waves.

leave it as an exercise for the reader to show that the situation in which the two roles are reversed yields the same answer. According to the principle of relativity, this must be the case—otherwise we would have a test that would enable us to tell whether we were at rest or moving uniformly. In Fig. 2 spherical light waves are shown emerging from the moving source.

We call the system at rest with respect to the moving source S′, and we call the system in which we are measuring the wave S. In S′ the frequency of the wave is ν', which means that once every T' seconds, as seen from S′, a light pulse is emitted. But S will claim (this is the meaning of Eq. 1.2) that the light is being emitted every T seconds where

$$T = \frac{T'}{(1 - v^2/c^2)^{1/2}} \tag{1.3}$$

This characteristic relativistic phenomenon is referred to as time dilatation. Let us suppose the source is moving away from the observer, then in a time interval ΔT, it will have moved a distance $v\,\Delta T$. This means that the total distance the light will travel between pulses is, according to S, who must take the time dilatation into account,

$$cT = \frac{vT'}{(1 - v^2/c^2)^{1/2}} + \frac{cT'}{(1 - v^2/c^2)^{1/2}} = T'\frac{v + c}{(1 - v^2/c^2)^{1/2}} = cT'\sqrt{\frac{1 + v/c}{1 - v/c}} \tag{1.4}$$

Remember how one derives the formula for the Doppler shift in classical physics. One argues that the extra distance the light travels when the source is in motion is $vT + cT$. Here we make the same argument, but we must also take into account the time dilatation. We have the relation

$$c\,\Delta T = \lambda \tag{1.5}$$

where λ is the wavelength of the light. Hence the stationary observer sees a "red-shifted" wavelength λ given by

$$\lambda = \lambda'\sqrt{\frac{1 + v/c}{1 - v/c}} \tag{1.6}$$

If the source is moving toward the observer, then in Eq. 1.6 we must replace v by $-v$ and there is a "blue shift."

The first application of red shifts to the motion of galaxies was made in the late 1920s by Edwin Hubble. The galaxies he studied were fairly slow-moving, so that he was able to use the familiar nonrelativistic approximation to Eq. 1.6:

$$\lambda \simeq \lambda' \left(1 + \frac{v}{c}\right) \tag{1.7}$$

However, recent observations of galaxies and quasars—objects a light-year or so in size that emit so much energy—10^{46}–10^{47} erg s^{-1} as compared to the sun's 10^{33} erg s^{-1} that they are thought to be powered by black holes—require use of the full relativistic formula, as we shall see later in the chapter. To appreciate Hubble's results we must now discuss the cosmic distance ladder, the ways in which distant astronomical objects are measured. This is also a subject to which one could devote an entire course. We will just give some of the flavor.

The reason why the cosmic distance ladder is called a "ladder" is that one uses the techniques that work at a smaller distance in order to move out additional steps (up the rungs) to farther and farther distances. We begin with the closest objects.

(a) For nearby astronomical bodies like the moon and the planets, the most accurate method for measuring distances is radar ranging. A radar signal is bounced off such an object, and the signal's roundtrip time is recorded. Thus, in the simplest approximation, the distance to one of these objects is given by

$$d = \frac{c}{2} \times t_{\text{roundtrip}} \tag{1.8}$$

Astronomical distance measuring, using radar signals, began in the 1960s. Listed here are a few of the planetary distances from the sun measured in what are called astronomical units (AU) where

$$1 \text{ AU} = 149{,}597{,}892.9 \pm 5.0 \text{ km} \tag{1.9}$$

Since planetary orbits are elliptical, the distances given are the semimajor axes of the orbits. Thus

Mercury, 0.387099
Earth, 1.00000
Saturn, 9.53884
Pluto, 39.44

The distances to the closer planets are now known with very high accuracy indeed.

(b) For stars out to about 30 pc we can use the method of parallax described in the first part of the book to measure the distances. The first parallax measurement ever made was done in 1838 by Friedrich Wilhelm Bessel at the Königsberg Observatory in Prussia. Curiously, he did this for 61 Cygni which is not the closest star. He found a parallax angle of $0.310'' \pm 0.02''$. The modern value is $0.292'' \pm 0.004''$. Using the results in the first part of the book, the reader can see what this corresponds to in miles or light-years. The star closest to us is α Centauri which has a parallax of 0.760″. This corresponds to a distance of 1.32 pc. The great virtue of the parallax method is that one does not have to know anything about the physics of the star. If one knows the intrinsic luminosity of a star—how much energy it puts out per second—finding its distance is relatively easy since the energy we receive from it falls off as the square of the distance from the emitter. (There are effects from the expanding universe that we will ignore in this chapter.) The problem is that when we observe a star, we cannot, simply by measuring its apparent brightness, tell if it appears very bright because it is close or whether it is farther away but intrinsically very luminous. The method of parallax avoids this problem altogether, but it cannot be used for stars farther away than about 30 pc, as the angles become too small to be measurable.

(c) For objects at a distance of between 30 and 200 pc, one can use something that is referred to as the convergent point method. This too avoids the luminosity problem. The principle is as follows.

Imagine that we have a distant object that moves in a circle with us at its center (Fig. 3).

If we could measure both the angular speed $\omega = \dot{\theta}$ and the tangential velocity v_t, then we would know the distance from the centre of the circle, d, through the relation

$$v_t = \omega d \tag{1.10}$$

The tangential velocities of the stars in a cluster like the Hyades are on the order of a few hundred kilometers a second. In the relativity theory there is a novel effect that depends on the dilatation of time (Eq. 1.3) known as the transverse Doppler shift. According to relativity, there is a Doppler shift even when an object is moving at right angles to the observer. But this is a purely relativistic effect, which means that it appears first to order $(v/c)^2$. Hence it is too small to be used to determine these stellar tangential velocities. But the radial velocities of stars—the velocities at which they move away from us—are amenable to Doppler shift measurements. Radial

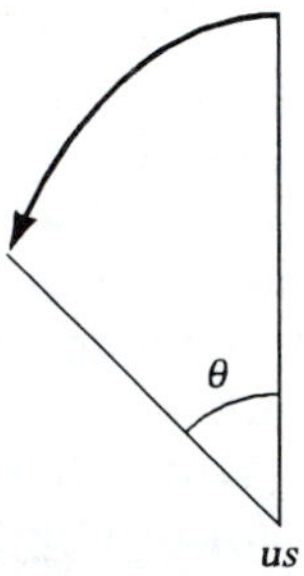

Figure 3 A distant object moving in a circle.

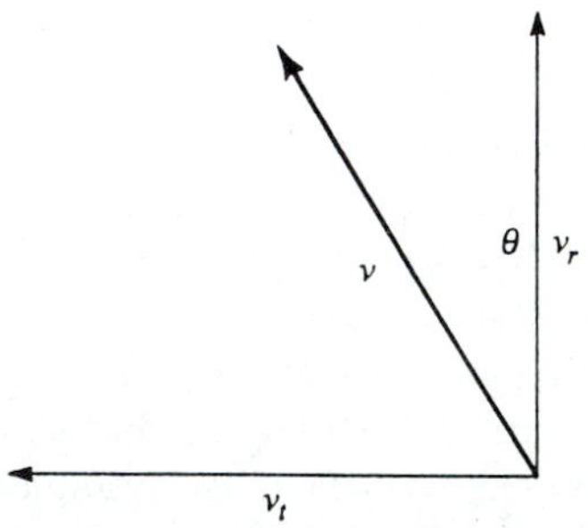

Figure 4 Vector diagram of the motion of a star in a cluster.

Doppler shifts are of order v/c. Indeed, that is what Hubble was measuring. Let us then draw a vector diagram of such a star in motion (Fig. 4).

If only we knew the direction of the stellar motion, we could find the tangential velocity by the relation

$$\tan\theta = \frac{v_t}{v_r} \tag{1.11}$$

Here astronomers make use of a trick of perspective. If a group of distant objects move parallel to each other, it appears to an observer as if they are moving toward a point. For example, when one looks at railroad tracks as they move off into the distance, it looks as if they are converging to a point. In this way, as stars move, astronomers measure the direction of **v**. Knowing the direction, they can use the relations

$$d = \frac{v_t}{v_r} \times \frac{v_r}{\omega} = \tan\theta \times \frac{v_r}{\omega} \tag{1.12}$$

to find d. Using this method, it was learned that the stars in the Hyades cluster are about 46 pc away. This is very important because knowing this distance and the apparent magnitude (the luminosity we perceive), we can determine the absolute magnitude (the amount of energy such a star actually emits). This serves as a checkpoint—a rung on the distance ladder—for methods that depend on knowing the luminosity. That is what we turn to next.

(d) Let us introduce some terminology. We call the absolute luminosity of a star L_A. It is measured in units of ergs per second. For example, the luminosity of the sun, $L_\odot$, is, as we have noted,

$$L_\odot = 3.826 \times 10^{33}\ \mathrm{erg\ s^{-1}} \tag{1.13}$$

The power P that an observer at a distance d receives from a star is, again ignoring effects of the expanding universe, related to L_A by the equation

$$P = \frac{1}{4\pi} \times \frac{1}{d^2} \times L_A \tag{1.14}$$

It is measured in units of erg $\mathrm{s^{-1}\ cm^{-2}}$. We can of course measure P. The problem is to determine L_A. Here is where the ladder comes in. Using the method of the previous section, we can measure the distance to various stars in the Hyades cluster. These stars, it turns out, belong to a variety of spectral types—have different colors. These spectral types are associated with stars having different intrinsic luminosities. We can determine this because we know their distances from us. Astronomers have invented a classification scheme for the spectral types, O, B, A, F, G, K, M, R, N, S, which they used to remember with the mnemonic "Oh, be a fine girl, kiss me right now, sweetheart," which reflects the era when nearly all astronomers were men. The O and B stars are the bluest, meaning they are the hottest—they have the highest surface temperatures—while the S stars are the coolest—they have the lowest surface temperatures. If the stars are "normal," in a sense we will discuss later, they lie on a plot known as the Hertzsprung-Russell diagram along what is known as the main sequence (Fig. 5).

The reason it is called a sequence is that, as stars age, they move along the diagram, taking up different positions sequentially. We shall come back to this later when we discuss the ages of astronomical objects. We can put the stars in the Hyades in their respective places on the diagram. But now we can consider the Pleiades star cluster whose distance we do not know. Its stars can also be put on the diagram. We have no reason to assume that they are not normal stars, so we can read off their luminosities and thus determine their distance—about 130 pc. This method works out to about 100,000 pc.

(e) The stars in the Hyades, as we have just seen, serve as "standard candles" to calibrate the luminosity of more distant stars. We can take the next step up the ladder, to about 4 Mpc, by using another kind of standard candle—stars with variable luminosities. In 1907 Henrietta Leavitt discovered a relationship between the period (the time between maxima in the luminosity curve) and the luminosity of a group of regularly pulsating stars in our galaxy called Cepheid variables. If one plots the luminosity as a function of the logarithm of the period, measured, say, in days, one finds almost a straight line; the longer the period, the brighter the star. In a

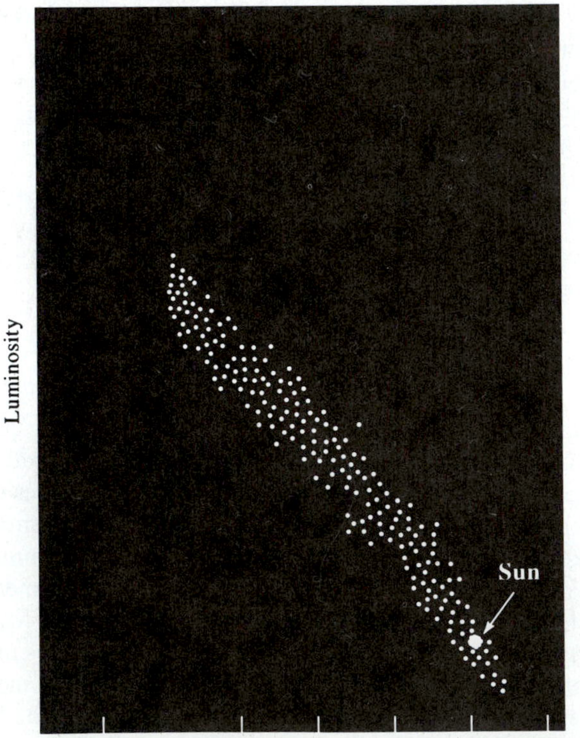

Figure 5 The Hertzsprung-Russell diagram showing only the main-sequence stars. We have plotted temperature as a function of luminosity. There are other ways of constructing the diagram such as using the power emitted versus the spectral type. The essential features remain the same.

general way, this is because brighter stars have bigger surface areas, and bigger objects oscillate more slowly than smaller ones. Since these stars are close enough so that we can measure their distances by one of the methods on the lower rungs of our ladder, we can determine their luminosities. But there are stars of similar variability in galaxies too distant for our other distance measures to work. Hence if we make the assumption that these variable stars obey the same kind of period luminosity relationship as the Cepheids, we can determine their distances. Astronomers argue about the validity of this assumption, so that there is some disagreement in that community about these distances. This uncertainty reflects itself, as we will see, in various places in cosmology. When distances get beyond several megaparsecs, the stars become too faint for us to be certain of their variability. However, for what we want to discuss next, the Hubble law of the expanding universe, the distance-measuring techniques described so far will do since the stars in Hubble's survey were relatively nearby.

In 1929 Hubble published a paper entitled, "A Relation Between Distance and Radial Velocity Among Extra-Galactic Nebulae." It is one of the most important scientific—let alone astronomical—papers ever published. For the purposes of his paper, Hubble did not include "nebulae" (galaxies) that were farther away than about 2 Mpc, so that one has, at least by present standards, some confidence in the distance measurements. Hubble measured the red shifts of these galaxies. It is important to understand what this means. Most of the visible light given off by a galaxy is in a spectral continuum like a rainbow (Fig. 6).

If this were the whole story, we could never detect a red shift since all that would happen would be that the continuum would be replenished as the violet and ultraviolet light was shifted to lower frequencies. The spectrum would look the same. But, fortunately, this is not the whole story. Cooler gasses circulate among the stars in a galaxy, and these gasses absorb light from the stars. This absorption is enhanced at certain frequencies. For example, some of these gasses absorb very strongly at the frequencies of the H and K lines of calcium—meaning that these gasses contain calcium. The effect of this absorption is to cut out bands in the visible spectrum and replace them by dark lines at these frequencies (Fig. 7).

These dark bands are related to specific transitions between the electron levels of atoms like calcium. When an astronomer is confronted by such a set of bands, the first thing that he or she tries to do is to associate them with specific electron energy levels in a plausible element. With distant galaxies these bands do not fall into

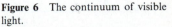

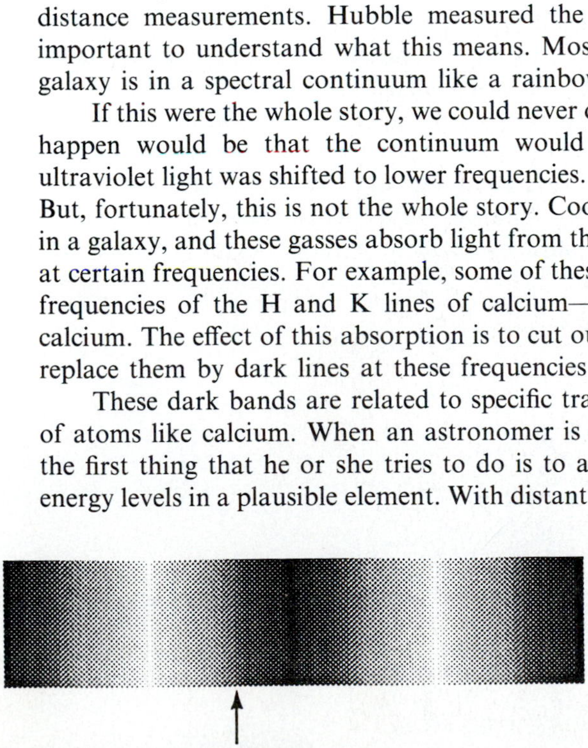

Figure 6 The continuum of visible light.

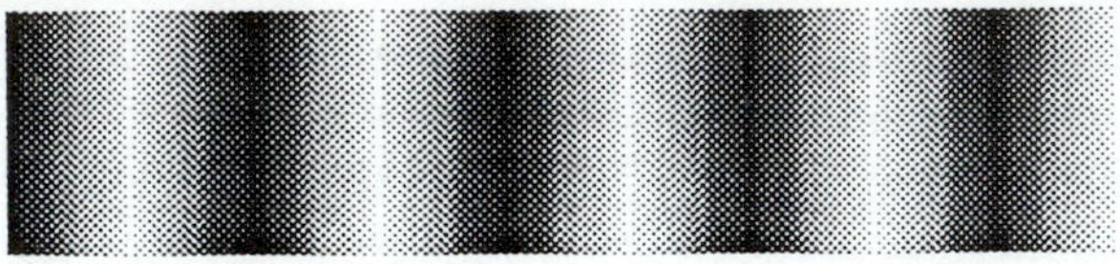

Figure 7 Dark bands in the absorption spectrum.

such a classification unless the assumption is made that the bands have shifted their frequencies. This is what is meant when one says that the red shift of a distant galaxy has been measured. As Fig. 8 shows, Hubble plotted the galaxies he observed as a curve relating the velocity of recession to the distance. To obtain the velocities he used the nonrelativistic Doppler shift formula, Eq. 1.7. The stars in his sample were sufficiently close so that one of the distance-measuring methods described above would be applicable.

There is a good deal of scatter in Hubble's original data, but nonetheless he concluded that there was a very simple relation between the recession velocity and the distance. They are simply proportional to each other, something that has become known as Hubble's law. There do not seem to be any blue-shifted galaxies. In modern notation Hubble's law is

$$v = H_0 d \tag{1.15}$$

We have introduced the constant of proportionality H_0 which is, needless to say, known as Hubble's constant. In fact, if the modern cosmological theories we are going to discuss are right, Hubble's constant isn't really a constant. It is a function of time. We have put the subscript zero on the H to indicate that it refers to the present time. The measurement of H_0 is one of the most important outstanding experimental problems in modern cosmology. It all comes down to improving the techniques for measuring distances—the sort of thing we were just discussing. We can sum up our ignorance in terms of a fudge factor h where, astronomers tell us,

$$0.4 \leqslant h \leqslant 1 \tag{1.16}$$

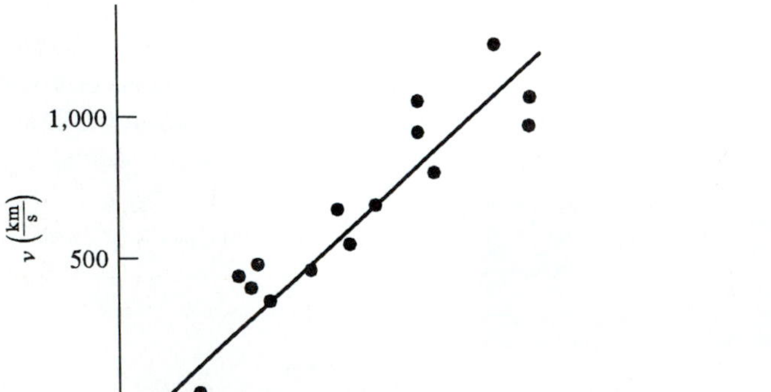

Figure 8 Hubble's plot of the velocity versus the distance to galaxies.

which means that, at the moment, we know h only up to a factor of 2. In terms of h we can write H_0 in a variety of ways:

$$H_0 = \frac{h \times 10^{-9}}{9.78\ \text{yr}} = h \times 3.24 \times 10^{-18}\ \text{s}^{-1} = h \times 100\ \text{km s}^{-1}\ \text{M pc}^{-1} \tag{1.17}$$

In the last form, which is the one most often used by cosmologists since it is easy to remember, the megaparsec is given in kilometers by

$$1\ \text{M pc} = 3.086 \times 10^{19}\ \text{km} \tag{1.18}$$

The inverse of H_0 is, roughly speaking, the "age of the universe"—the time during which the universe has been expanding. We shall make this connection more precise later. Note that

$$\frac{1}{H_0} = 9.780 \times 10^9\ \text{yr h}^{-1} \tag{1.19}$$

Recall from the Micropaedia that the ages of things we know about, like the moon and the planets, is about 4.5×10^9 yr. This tells us that the upper bound on h is pretty close to what can be tolerated.

Cosmologists have introduced a quantity Z to measure the percentage of red shift. It occurs ubiquitously in papers and books on the subject and is defined in terms of the frequency v_{em} emitted by, say, a galaxy and the frequency v_{obs} observed by us. The difference between these two frequencies comes about because of the expansion of the universe—the red shift. Thus

$$Z = \frac{v_{\text{em}}}{v_{\text{obs}}} - 1 = \frac{\lambda_{\text{obs}}}{\lambda_{\text{em}}} - 1 = \sqrt{\frac{1 + v/c}{1 - v/c}} - 1 \simeq \frac{v}{c} \tag{1.20}$$

Therefore, for slowly moving galaxies, Z is a direct measure of the recession velocities. In general, if we know Z, we can invert to find v/c; i.e.,

$$\frac{v}{c} = \frac{(1+Z)^2 - 1}{(1+Z)^2 + 1} \tag{1.21}$$

Note that as Z tends to infinity, v tends to c. The present record holder for Z is a quasar with $Z = 4.9$. This corresponds to a v/c of 0.94. These distant quasars are receding with relativistic speeds. It is important to remark that this connection between Z and velocity ignores the fact that space may be "curved" and is expanding. In a later chapter we will introduce these corrections. We will see that they depend on what model of cosmology one assumes. The results we have given here assume a model in which space is flat; i.e., the geometry of light rays is Euclidean.

Let us naively apply the Hubble law to determine the present size of the universe, R_U. Neglecting relativistic effects, we can write

$$R_U = \frac{v}{H_0} = \frac{v}{c} \times \frac{c}{H_0} \tag{1.22}$$

If we put in the v/c of the most distant thing we see, the quasar with $Z = 4.9$, we find that R_U, is given by

$$R_U \simeq 10^{28}\ \mathrm{cm\ h^{-1}} \tag{1.23}$$

This was the basis for the estimate of the size of the visible universe quoted in the Micropaedia. We next turn to the question of what the Hubble law means.

The alert reader will have no doubt noticed that Hubble's law does not contain an angle. The recession velocity at a given distance is the same in all directions. Figure 9 is a vector diagram of the situation.

One might greet this situation with a certain delusion of grandeur, thinking that we are at the center of it all and that the galaxies are recoiling from *us*. It is much more reasonable to imagine a model in which all the galaxies are moving away from each other and in which an observer on any given galaxy would discover the same Hubble law. If we could find some way of synchronizing the times, all such observers would presumably agree about the value of Hubble's constant at that time. A homey model for this is to think of the galaxies as if they were glued to the surface of a balloon that is being blown up. Then, each spot on the balloon would recede from the others uniformly as the balloon is inflated uniformly (Fig. 10).

In cosmology the force that runs this expansion is gravitation. The best theory we have of gravitation—although one that is limited by the fact that it has been so difficult to incorporate quantum mechanics into it—is Einstein's general theory of relativity. This is not an easy theory to learn because of the rather advanced mathematics it involves. The goal of this book is to familiarize the reader with the general features of cosmology without introducing the full machinery of this theory. As we will see, it is surprising how far we can go with this program. But it would be unfortunate if we did not at least give the reader some of the flavor of Einstein's

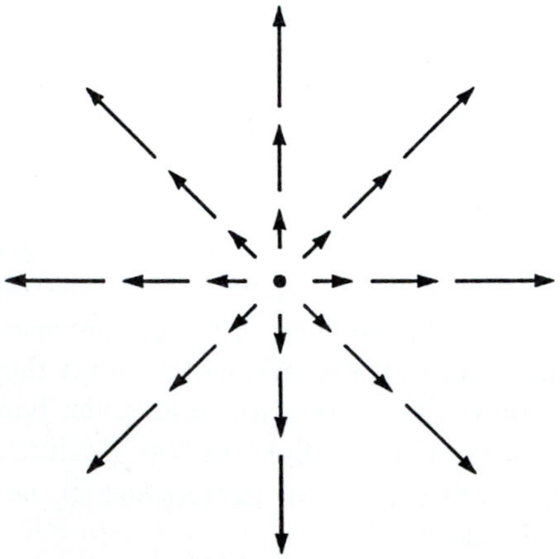

Figure 9 A vector diagram of the galaxies receding from us with their radial velocities increasing with distance.

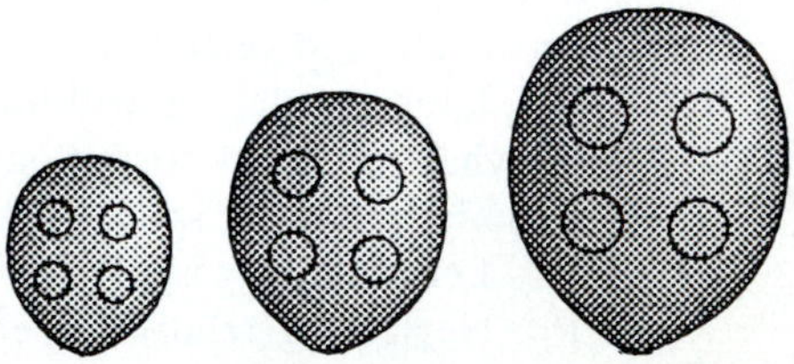

Figure 10 Galaxies on the surface of a balloon that is inflating in three stages.

magnificent edifice. That is what we shall do now, and then we shall explain why we do not need the whole structure. Two basic results of Einstein's theory are

1. Gravity alters clocks.
2. Gravity "curves" space.

Let us deal with the second result first. To make things as concrete as possible, let us imagine that we could make a huge triangle—megaparsecs in size—out of light rays in space (Fig. 11).

We may now imagine measuring the angles of this triangle. If we are open-minded about this, we do not insist a priori that the interior angle sum of the triangle is 180°. It is a question for experiment to determine what the angle sum really is. In fact, this question was first raised nearly two centuries ago by mathematicians like Johann K. F. Gauss and later Bernhard Riemann. They raised it in the context of so-called non-Euclidean geometries. The reader may remember how one proves that the angle sum of a triangle is 180°. One uses the postulate of parallels; i.e., through a point external to a given straight line one and only one parallel straight line can be drawn. For centuries mathematicians struggled to "prove" this postulate but finally gave up when they learned that perfectly consistent geometries could be constructed if, for example, one assumed that through such an external point no such parallel lines can be drawn. A realization of such a geometry consists of the geometry made out of segments of great circles on the surface of a sphere. In fact, if one makes up a triangle out of such great circular segments, one will find that its angle sum is greater than 180°. A surface where this happens is said to have "positive curvature." Likewise, if one draws the curves of the shortest distance between two points—what one means in general by a straight line—on the surface of a saddle and constructs a triangle out of these lines, one will find that its angle sum is less than 180° which one characterizes by saying that the saddle space has "negative curvature." But what has this to do with gravitation?

This is where Einstein's genius came in. His first step was to look at Newton's law of gravitation in a new way. We would write Newton's force law in the form

$$mMG\frac{\mathbf{r}}{r^3} = m\ddot{\mathbf{r}} \tag{1.24}$$

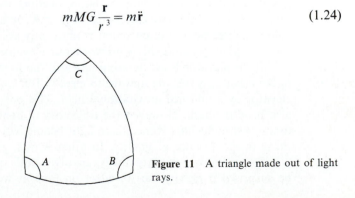

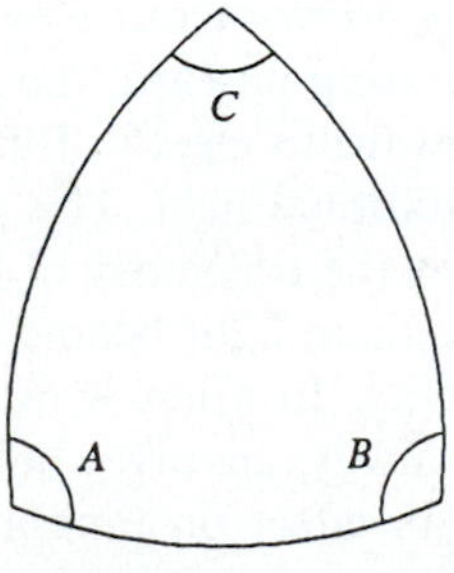

Figure 11 A triangle made out of light rays.

Unless we were Einstein, it would not occur to us that there was anything remarkable about this familiar equation. But he noticed that the symbol m was being used here in two quite distinct ways. On the left side, m is a measure of the strength of the gravitational coupling between the two masses m and M. On the right side, however, m is a measure of the inertia of the object—how difficult it is to accelerate. The left-hand m we may call, following Einstein, the "gravitational mass," m_{gr}, while the right-hand m we may call the "inertial mass" m_{in}. If we think about it, we will realize that we have always assumed that these two masses are the same and have happily canceled them out on the two sides of Eq. 1.24. This leads to the conclusion that in a uniform gravitational field all objects—irrespective of mass—fall with the same constant acceleration.

Early in this century it actually occurred to the Hungarian baron Roland Eötvos that the equivalence of these two masses should be tested experimentally. He used a torsion balance and found that

$$\frac{m_{in}}{m_{gr}} = 1 + O(10^{-8})$$

Using more modern methods, experimenters have improved this result so that now

$$\frac{m_{in}}{m_{gr}} = 1 + O(10^{-11})$$

Even before Einstein had heard of Baron Eötvos's result, he had decided to elevate the equivalence of these two masses to a principle, the principle of equivalence; i.e.,

$$\frac{m_{in}}{m_{gr}} = 1$$

But this, he realized, entailed the following remarkable consequence. Let us imagine that we are inside an elevator in outer space, away from all masses, and that it is being pulled up by a rope at a constant acceleration g. If we drop an object, any object, the floor will accelerate toward it with this acceleration. But we can describe this situation equivalently by imagining that the elevator is at rest in a uniform gravitational field just strong enough to produce the same acceleration on all falling objects. Conceptually we can switch back and forth between a uniform gravitational field and a frame of reference that is uniformly accelerating. Now suppose someone shines a light ray into the window of our elevator (Fig. 12).

If the elevator is accelerating upward, the light will appear bent since the floor of the elevator is rushing upward to meet it. But we can now replace the accelerating elevator by a uniform gravitational field. This will, by the principle of equivalence, also bend the light. However, the trajectory of light in a vacuum is the path of least time—a straight line. Hence these light beams will define a different geometry than those in the absence of gravity. In other words, gravity has curved space.

As we know, relativity always involves both space and time. So we should not be surprised if gravity has an effect on time as well as space. To see this, we can

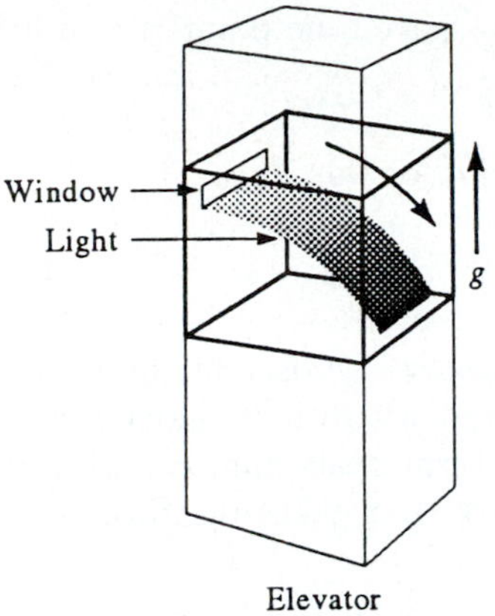

Figure 12 Light traversing an accelerating elevator.

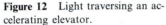

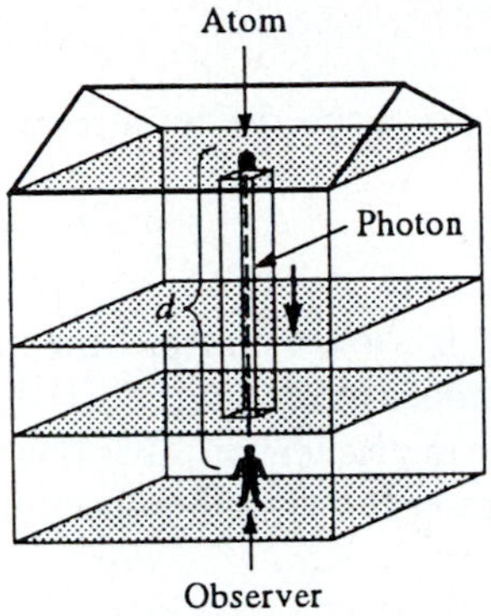

Figure 13 Light falling a distance d in a uniform gravitational field.

consider the following setup. We have an observer in the basement and an atom in the attic that is emitting light down a shaft to the basement (Fig. 13).

We can think of the light as being made up of photons being accelerated in the gravitational field with an acceleration g. But this field can be replaced by an observer accelerating upward toward the atom with the same acceleration. To order v/c it takes a time d/c for the light to meet the observer. By this time, the observer is traveling with a speed v given by

$$v = g \times \frac{d}{c} \tag{1.25}$$

Hence the light will appear blue-shifted by an amount

$$\nu' = \nu\left(1 + g \times \frac{d}{c}\right) \tag{1.26}$$

This means that light entering a gravitational field will be blue-shifted, while light escaping from a gravitational field will be red-shifted. But these frequencies are a kind of clock, which is the basis for the statement that gravity alters the behavior of clocks.

In nature, there are no precisely uniform gravitational fields, so that the principle of equivalence has only limited applicability in solving the problems of gravitation. After Einstein formulated it he went on to create the full general theory of relativity and gravitation. The essential idea of this theory is that at every point in space-time there is a metric—a geometry—that is determined by the gravitational fields at that point. Whether we use the term or not, we are all familiar with metrics. For example, using the Pythagorean theorem, we can write the distance between two points in ordinary three-dimensional space as

$$s^2 = x^2 + y^2 + z^2 \tag{1.27}$$

This is the so-called Euclidean metric. In the special theory of relativity this becomes generalized to the distance between two space-time points:

$$s^2 = x^2 + y^2 + z^2 - (ct)^2 \tag{1.28}$$

We can write the infinitesimal version of this as

$$ds^2 = \sum_{\mu,\nu=1}^{4} dx^\mu \, dx^\nu \, L_{\mu\nu} \tag{1.29}$$

where $L_{\mu\nu}$ is a 4×4 matrix—the metric tensor—whose only nonzero components are the diagonal components 1, 1, 1, -1 and where it is understood that $dx^4 = c\,dt$. When the metric tensor takes this simple form, space-time is said to be flat. However, in the presence of gravitational fields, Eq. 1.29 takes the form

$$ds^2 = \sum_{\mu,\nu=1}^{4} dx^\mu \, dx^\nu \, g_{\mu\nu} \tag{1.30}$$

where $g_{\mu\nu}$, the metric tensor, is a function of space and time. There is no reason to believe that, in general, the off-diagonal matrix elements of $g_{\mu\nu}$ will be small, so that in strong gravitational fields space and time are mixed up together. It takes the full machinery of Einstein's theory to determine $g_{\mu\nu}$ when gravitational fields are present but, fortunately, we can give a decent account of much of cosmology without using the full formalism.

The reason is that, to a good approximation, the galaxies, the cosmic background radiation, and even the putative cosmic neutrinos are distributed, on the average, uniformly in the universe. Of course, there are ripples and lumps—like the earth—but these average out in the grand scheme. This means that at any given time we can take the matter density ρ_m to be independent of position. This simplifies life enormously. To see why, we remind the reader of the following problem in Newtonian gravitation. We imagine a huge space filled uniformly with matter and then hollow out a sphere somewhere in the space. We will now argue that within this sphere there is no gravitational field, something that depends, as we shall see, both on our assumption of a uniform distribution of matter and the fact that the gravitational force law has a $1/r^2$ character. Figure 14 illustrates the situation. The sphere has radius R, and P is an arbitrary point within it.

Let us call the total mass in a shell of thickness dr, on the surface of the sphere of radius R, M. Since this mass is uniformly distributed, the amount of mass on the fraction of the shell with area $2\pi R^2 \sin\theta \, d\theta$, M_S, is given by

$$M_S = \frac{2\pi R^2 \sin\theta \, d\theta}{4\pi R^2} M = \frac{\sin\theta \, d\theta}{2} M \tag{1.31}$$

This acts on a mass m located at an arbitrary point P within the sphere. Thus the gravitational potential at P due to this shell, V_P, is given by

$$V_P = -\frac{mMG}{2} \int_0^\pi \frac{\sin\theta}{r} \, d\theta \tag{1.32}$$

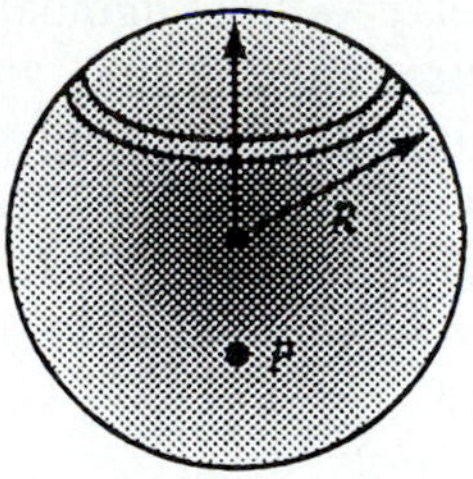

Figure 14 A hollowed-out sphere of radius R in a uniform distribution of matter. Also shown is a spherical shell on the surface of the sphere with area $2\pi R^2 \sin\theta\, d\theta$.

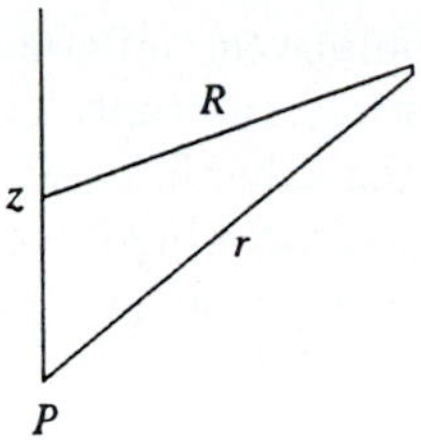

Figure 15 The coordinates in the integration.

In Fig. 15 we show the coordinates R, r, and z, where z is the distance from P to the center of the sphere.

We have the relation

$$r^2 = R^2 + z^2 + 2Rz\cos\theta \tag{1.33}$$

or

$$\sin\theta\, d\theta = -\frac{r\,dr}{Rz} \tag{1.34}$$

Hence we can write

$$V_P = \frac{mMG}{2}\int_{r+z}^{r-z} \frac{1}{R}z\,dr = -\frac{mMG}{R} \tag{1.35}$$

The dependence of the potential on z has dropped out, so that the potential due to this spherical shell is independent of the point P within the sphere. It is the same at all points, which means that there is no force. This is the contribution from a single shell, but clearly we can add up the shells one after the other to reach the conclusion that the external mass does not produce a force within the hollowed-out sphere. But the hollowed-out sphere is of any size anywhere in this essentially infinite space. Hence a test particle in this space will feel no gravitational force.

It is perhaps not surprising that there is a similar result in the general theory of relativity. It says essentially that the geometry generated by such a uniform distribution of matter is flat. We are free then to use the ideas of Newtonian mechanics and, when needed, the special theory of relativity, to see us through. Before we get too carried away by this prospect, let us demonstrate that the notion that matter is distributed uniformly in infinite space, independently of time, leads to a puzzle. It goes under the name of the Olber's paradox, although many people seem to have had a hand in its formulation. Let us call the number of stars per centimeter cubed in this putatively infinite volume n and let us assume n is a constant independent of space and time. Let us also assume that all these stars have, to first approximation, the same intrinsic luminosity L. This is certainly wrong, but to get out of the paradox we would need to give L a very unnatural dependence on the distances of the stars from us. Thus the number of stars in a spherical shell at a distance r from us, $N(r)$, is

$$N(r) = n 4\pi r^2 \, dr \tag{1.36}$$

But at a distance r, the power received from a given star varies as $1/r^2$. Therefore, according to this model, the total power P from all the stars would be

$$P = Ln \int_0^\infty dr = \infty \tag{1.37}$$

This is of course a crazy result. We will discuss the resolution of this dilemma when we see in more detail how the theory of the expanding universe works. In any event, the reader should now have some idea of why observationally we think the universe is expanding. Next we turn to the theory.

2

The Expanding Universe—Theory

In the previous section we presented some of the observational arguments as to why we believe that the universe is expanding. In this section we will describe some of the theoretical models that are consistent with what we observe. We begin with two basic postulates that summarize our experimental information.

1. On the average, the universe is at present isotropic. This means that any observer would see the same general features no matter in which direction he or she looked.
2. On the average, the universe is at present homogeneous. This means that an observer stationed anywhere in it would see about the same thing as any other observer.

If we make the assumption that these postulates held as the universe was expanding in the past, we can derive Hubble's law, which states that the expansion velocity is proportional to the distance from an observer. In making this argument we shall assume that space is flat, although much of what we are going to discuss does not depend on that. Later in the book we shall worry about how the curvature of space affects phenomena like the propagation of light. Let us imagine a triangle at whose vertices are three galaxies (Fig. 1).

The assumption that the universe remains homogeneous and isotropic as it expands implies that this triangle will expand into a similar one (Fig. 2). If $l(t)$ is the length of a side—we might as well make the triangles equilateral—at time t, then, in order to preserve self-similarity, $l(t)$ must be related to $l(0)$ by the equation

$$l(t) = R(t)l(0) \tag{2.1}$$

We can now imagine covering the universe with a mesh of such triangles (Fig. 3).

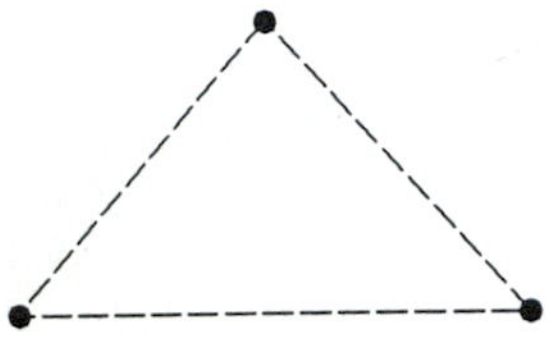

Figure 1 A triangle with three galaxies at its vertices.

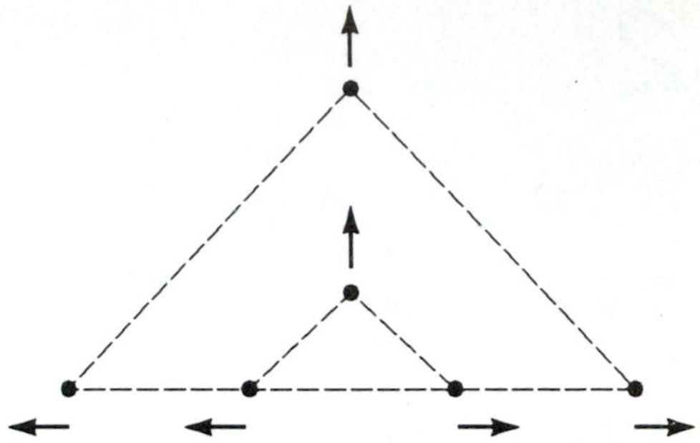

Figure 2 The expansion of triangles into similar ones.

If homogeneity and isotropy of the universe as a whole are to be preserved, the scale factor $R(t)$ must be universal; i.e., it cannot depend on position. But notice (the overdot always stands for the time derivative)

$$\dot{l}(t) = \dot{R}(t)l(0) = l(t)\frac{\dot{R}(t)}{R(t)} = l(t)H(t) \tag{2.2}$$

where

$$\frac{\dot{R}(t)}{R(t)} = H(t)$$

This is Hubble's law if we identify $H(t)$ with Hubble's "constant". We may now connect this expansion to the red shift in the following way. Let us consider an observer sitting on a galaxy at rest with respect to the Hubble flow. In our balloon analogy such an observer would be resting on one of the patches painted on the surface of the balloon. Such an observer is said to be "comoving" with respect to the Hubble flow. Now we can imagine a particle that moves past this observer. Suppose the instantaneous velocity of the particle with respect to the observer at time t is $v_1(t)$. In a tiny interval dt the particle will have moved a distance $v_1(t)\,dt$. What speed—call it $v_2(t+dt)$—will a comoving observer at this distance from the first observer assign to the particle? If the universe were not expanding, this would be a nonquestion since the velocity would be the same. But according to Hubble's law, the second observer is receding from the first with a speed v_{rec} given by

$$v_{\text{rec}} = v_1(t)\,dt\,\frac{\dot{R}(t)}{R(t)} \tag{2.3}$$

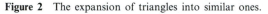

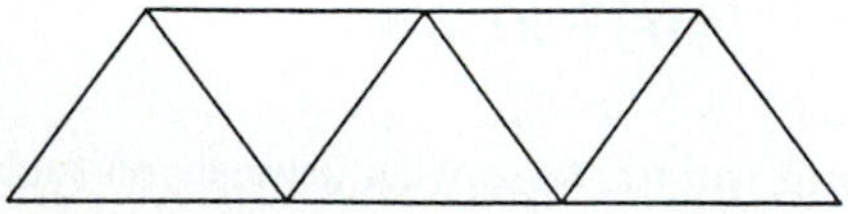

Figure 3 The universe covered with a mesh of triangles.

Hence $v_2(t + dt)$ will be given by

$$v_2(t + dt) = v_1(t) - v_1(t)\, dt \frac{\dot{R}(t)}{R(t)} = v_1(t) + \dot{v}_1(t)\, dt \tag{2.4}$$

or, dropping the subscripts,

$$\frac{\dot{v}(t)}{v(t)} = -\frac{\dot{R}(t)}{R(t)} \tag{2.5}$$

Thus

$$v(t) = \frac{v(0)}{R(t)} \tag{2.6}$$

This means that a particle that has been propagating toward a comoving observer from the distant past, while the universe has been expanding, will appear to have slowed down as compared to the velocity it would have had if the universe had not been expanding.

We can now apply these ideas to the propagation of light. In the interval dt, light will have traveled a distance $c\, dt$, so that, according to Hubble's law,

$$v_{\text{rec}} = c\, dt \frac{\dot{R}(t)}{R(t)} \tag{2.7}$$

where the speed of light c is not a function of time. Thus, to this observer, light will appear Doppler-shifted to the red by an amount

$$d\lambda = \lambda c\, dt \frac{\dot{R}(t)}{R(t)} \tag{2.8}$$

where we have used the nonrelativistic Doppler shift formula. Therefore

$$\lambda(t) = \lambda(0) \frac{R(t)}{R(0)} \tag{2.9}$$

In other words, light propagating toward us from the distant past, as the universe is expanding, will appear red-shifted as compared to the wavelength it would have had if the universe had not been expanding. Put another way, the momentum of the photons, h/λ, decreases as $1/R(t)$. It is customary to restate this result in terms of the parameter Z, really $Z(t)$, introduced in the last section. Thus

$$Z(t) = \frac{\lambda(t)}{\lambda(0)} - 1 \tag{2.10}$$

or, using Eq. 2.9,

$$1 + Z(t) = \frac{R(t)}{R(0)} \tag{2.11}$$

It is clear then that the expansion of our homogeneous, isotropic universe is governed by a single universal function $R(t)$. We turn next to how this function is determined.

We recall our discussion in the last chapter of the hollowed-out sphere in a uniform matter distribution. If we fill our sphere with a uniform matter distribution corresponding to a mass M, it will act on a mass point m at a distance $R(t)$ from the center as if the entire mass M were concentrated at the center of the sphere (Fig. 4).

Newton's law tells us that

$$m\ddot{R}(t) = -\frac{GmM}{R(t)^2} \tag{2.12}$$

Thus multiplying both sides of Eq. 2.12 by $\dot{R}(t)$, we have

$$\frac{d}{dt}\frac{\dot{R}(t)^2}{2} = \frac{d}{dt}\frac{GM}{R(t)} \tag{2.13}$$

We can integrate this to find

$$\frac{\dot{R}(t)^2}{2} - \frac{GM}{R(t)} = E \tag{2.14}$$

where E is an integration constant with the dimensions of energy. In fact, the first term on the left-hand side of Eq. 2.14 can be interpreted as the kinetic energy, and the second term as the potential energy. Speaking of dimensions, the $R(t)$ in this equation has dimensions of length. It doesn't matter what length, just some convenient reference length. In Eq. 2.1, for example, this length scale has been factored out so that the $R(t)$ in that equation is dimensionless. We have used the same symbol in both instances to avoid more notation. For later purposes it is useful to introduce the mass density $\rho_M(t)$. This can be, according to our assumptions, a function of time but not of space. It is related to the total mass M by the equation

$$M = \tfrac{4}{3}\pi \times R(t)^3 \times \rho_M(t) \tag{2.15}$$

Thus Eq. 2.14 becomes

$$\frac{\dot{R}(t)^2}{2} - \frac{4\pi}{3} \times GR(t)^2 \times \rho_M(t) = E \tag{2.16}$$

We can now draw on our knowledge of Newtonian mechanics to classify the solutions of Eq. 2.14.

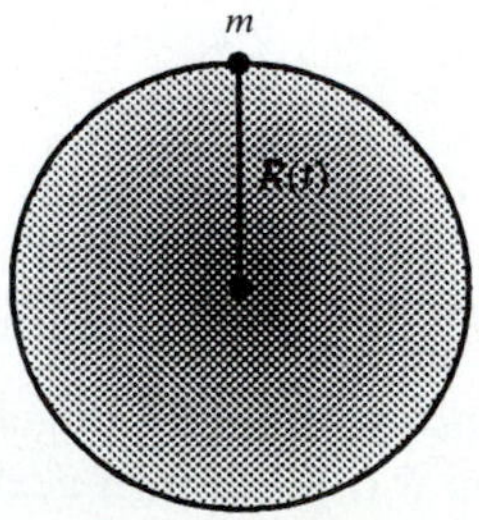

Figure 4 A mass point m on the surface of a sphere of radius $R(t)$.

a. $E < 0$. In this case the potential energy term, which according to Eq. 2.14 is proportional to $1/R(t)$, dominates. Thus $R(t)$ cannot increase without limit. In the Newtonian case this means that the massive object is trapped in a bound orbit. In the case at hand, it means that, after some finite time, the universe will stop expanding.
b. $E = 0$. This is the borderline between positive and negative E. In this case the universe slows down as it expands and comes to a stop at $t = \infty$. In the Newtonian analog this means that the massive object has just enough energy to escape. Because of its simplicity, this is the situation that most theorists hope is the one that nature has chosen. It has the property that the matter density—more generally the energy density—is fixed in terms of the Hubble constant. This special density is usually called the critical density $\rho_c(t)$. From Eq. 2.16,

$$\rho_c(t) = \frac{H(t)^2 3}{8\pi G} = \frac{3}{8\pi} \times \frac{H(t)^2}{G} \tag{2.17}$$

It is customary to rewrite ρ_c in terms of the Planck mass M_P using the relationship we derived in the first section of the book:

$$G = \frac{\hbar c}{M_P^2} \tag{2.18}$$

Thus

$$\rho_c(t) = \frac{3}{8\pi} \times \frac{M_P^2}{\hbar c} \times H(t)^2 \tag{2.19}$$

If we put in the value for $H(t)$ for the present universe and call the present critical density simply ρ_c, we have, recalling from the previous chapter the definition of the fudge factor h (Eq. 1.16),

$$\rho_c = 1.879 \times 10^{-29} h^2 \text{ g cm}^{-3} \tag{2.20}$$

We will have much occasion to return to the question of whether or not the actual density of the universe is ρ_c.
c. $E > 0$. The universe expands forever. Some people find this a gloomy prospect since it presumably means that the entire universe will eventually undergo a kind of heat death.

In addition to massive matter we also want to include the mass equivalent of the energy density of massless objects like the cosmic background photons in our equation for $R(t)$. To this end we shall call the full density ρ, leaving the designation ρ_M for that part of the density that comes from massive matter such as neutrons and protons. We can always rescale $R(t)$ so that we can write Eq. 2.16 in the form

$$\dot{R}(t)^2 + kc^2 = \frac{8\pi}{3} G\rho(t) R(t)^2 \tag{2.21}$$

We have the correspondence

$$E < 0 \leftrightarrow k = 1$$

$$E = 0 \leftrightarrow k = 0$$

$$E > 0 \leftrightarrow k = -1$$

but we should not lose sight of the fact that the E in the original equation can have any value.

We can use Eq. 2.21 to give, within the context of this model, a precise definition of the age of the universe which we call t_0. Thus

$$t_0 = \int_0^{t_0} dt = \int_0^{R_0} \frac{dR}{\dot{R}} = \int_0^{R_0} \frac{1}{[(8\pi/3)\rho R^2 - c^2 k]^{1/2}} dR \tag{2.22}$$

But k is a constant, so that we can solve for it in terms of the present values of the variables, which we denote with a subscript zero:

$$\frac{8\pi G}{3} \times \rho_0 R_0^2 - \dot{R}_0^2 = kc^2 \tag{2.23}$$

Notice the following relation,

$$\left(\dot{R}_0^2 - \frac{8\pi G}{3} \times \rho_0 R_0^2 + \frac{8\pi}{3} G \times \rho R^2\right)^{1/2} = R_0 H_0 \left(1 - \frac{\rho_0}{\rho_c} + \frac{\rho}{\rho_c} \times \frac{R^2}{R_0^2}\right)^{1/2} \tag{2.24}$$

If we call $x = R/R_0$, we have, finally, the elegant expression

$$H_0 t_0 = \int_0^1 \frac{1}{\left(1 - \frac{\rho_0}{\rho_c} + \frac{\rho}{\rho_c} x^2\right)^{1/2}} dx \tag{2.25}$$

There is not much more that we can say about this expression without knowing the disposition of ρ. To make progress, we shall use some results that we will argue for in detail later in the book. If present ideas are right, from about 100,000 yr after the Big Bang until the present, the universe has been what cosmologists call "matter-dominated". This means that the density ρ has been dominated by ρ_M for most of the approximately 15 billion years since the Big Bang. In addition to the neutrons and protons that we know about, there might be other forms of massive matter. For example, the neutrinos might have small masses. In any event these particles do not disappear in the course of time. Let us call the number of the ith species N_i, its mass m_i, and the volume of space V. Then we can write ρ_M as

$$\rho_M = \sum_{i=1}^{n} \frac{m_i N_i}{V} \tag{2.26}$$

But since the length scales as $R(t)$, the volume V must scale as $R(t)^3$. Thus

$$\frac{\rho_M}{\rho_{M0}} = \left[\frac{R(0)}{R(t)}\right]^3 \tag{2.27}$$

Hence in the matter-dominated regime—most of the life of the universe—we have the even more elegant formula

$$H_0 t_0 = \int_0^1 \frac{1}{\left[1 + \frac{\rho_0}{\rho_c}\left(\frac{1}{x} - 1\right)\right]^{1/2}} dx \tag{2.28}$$

Since ρ_0/ρ_c is positive and since $1/x$ is, in the range of integration we are considering, greater than or equal to 1, we have

$$\left[1 + \frac{\rho_0}{\rho_c}\left(\frac{1}{x} - 1\right)\right]^{1/2} > 1 \tag{2.29}$$

Thus

$$H_0 t_0 < 1 \tag{2.30}$$

or

$$t_0 < \frac{1}{H_0} = 9.78 \times 10^9 \text{ yr h}^{-1} \tag{2.31}$$

This is the more precise version of the discussion we presented in connection with Eq. 1.19 of the previous section.

The most agreeable case to work out in detail is case b, that is, $E = k = 0$. It is customary to define the quantity $\Omega(t)$ by

$$\Omega(t) \doteq \frac{\rho(t)}{\rho_c(t)} \tag{2.32}$$

This case then corresponds to $\Omega = 1$. This is a time-independent statement. If the universe begins its expansion with its density at precisely the critical density, that condition will maintain itself at all times. We may therefore ask how close we are to the critical density in the present universe. This is one of the most important unanswered experimental questions in modern cosmology. The problem is that we don't know how much the dark matter contributes. We know, as we have explained earlier, that there must be a great deal of dark matter, but what is it? What is its collective mass? We are, in a manner of speaking, still in the dark. There are, however, two statements we can make. We can measure a quantity we call Ω_γ; that is,

$$\Omega_\gamma = \frac{\rho_\gamma}{\rho_c} \simeq 10^{-5} \text{ h}^{-2} \tag{2.33}$$

This is clearly very far from unity. On the other hand, we can also measure a quantity we call Ω_M; that is, $\Omega_M = \rho_M/\rho_c$. There are various ways of trying to measure Ω_M. If we measure only the matter we can see, $\Omega_M \simeq 0.002$. But if we include the baryonic dark matter which the dynamics of galaxies indicates must be there, Ω_M is about two orders of magnitude larger. In our chapter on helium production we shall point out that cosmological nucleosynthesis can be used to set an upper bound to it; i.e.,

$$\Omega_M < \frac{0.019}{h^2} \leqslant 0.12 \tag{2.34}$$

where we have taken h at its minimum of 0.4 to get the last inequality. This is tantalizingly close to unity. One wonders if nature is perverse enough to produce such a close approximate value without taking advantage of the really elegant possibility of setting $\Omega = 1$.

For the $\Omega = 1$ case we have the exact relation, for matter dominance, of

$$H_0 t_0 = \int_0^1 x^{1/2}\, dx = \frac{2}{3} \tag{2.35}$$

so that

$$t_0 = \frac{2}{3H_0} = 6.46 \times 10^9 \text{ yr h}^{-1} \tag{2.36}$$

which, with the lower values of h, can lead to a very acceptable age for the universe.

For the case $k = 0$ we can solve Eq. 2.21 explicitly. For the matter-dominated situation, for example, we have

$$\dot{R}(t)^2 = \frac{\text{constant}}{R(t)} \tag{2.37}$$

We can try a solution of the form $R(t) \sim t^n$ and substitute in and solve for n. If we do, we find at once that $n = \frac{2}{3}$. Thus, in this case, we have the very important result that

$$R(t) = \text{constant} \times t^{2/3} \tag{2.38}$$

so the universe is expanding according to a simple power law. Note further that, in general,

$$\ddot{R}(t) = -\frac{GM}{R(t)^2} \tag{2.39}$$

so that

$$\ddot{R}(t) < 0$$

But, in the expanding phase, $\dot{R}(t) > 0$. In other words, for this model, $R(t)$ is a convex function. In Fig. 5 we present a plot of it. For comparative purposes, it is useful to consider the unphysical case in which $\rho = 0$. In this case we can deduce that

$$R(t) = R(t_0)\frac{t}{t_0} \tag{2.40}$$

so that

$$\frac{\dot{R}(t_0)}{R(t_0)} = \frac{1}{t_0} \tag{2.41}$$

This unphysical limiting case then satisfies

$$H_0 t_0 = 1 \tag{2.42}$$

In Fig. 5 we plot both the matter-dominated case and the unphysical case for $k = 0$.

For completeness we now wish to describe the radiation-dominated regime—the first 100,000 yr. Once again we assume that the student has some familiarity with the

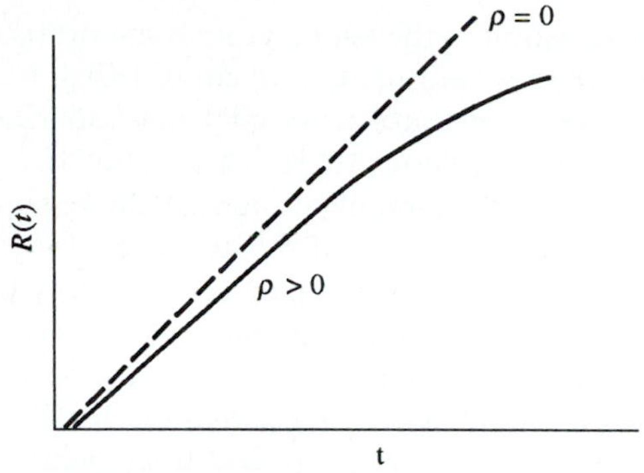

Figure 5 A plot of $R(t)$ for the matter-dominated $k = 0$ case along with a plot of the unphysical zero mass density case for comparison.

theory of blackbody radiation. The result that we need here is that in a blackbody at temperature T the number of photons per cubic centimeter, n_γ, is proportional to T^3. In fact, in Eq. 1.46 of the Micropaedia we presented the exact formula, i.e.,

$$n_\gamma = \left(\frac{kT}{\hbar c}\right)^3 \frac{2.4}{\pi^2} \tag{2.43}$$

But n_γ is proportional to $1/V$, where V is the volume which is in turn proportional to R^3. For this to be consistent with Eq. 2.53 we must have $T \sim 1/R$. So long as there is no mechanism for creating photons, we must have the photon temperature falling off as $1/R$. As we shall see, there are mechanisms for creating photons such as particle–antiparticle annihilation into photons. When these occur, there are jumps in the photon temperature that we can and will calculate. We shall neglect such mechanisms here and content ourselves with finding how R depends on time in the radiation-dominated regime. To this end we note from the dependence of T on R that

$$\frac{\dot{T}(t)}{T(t)} = -\frac{\dot{R}(t)}{R(t)} \tag{2.44}$$

But, as we saw in the discussion of Eq. 1.46 of the Micropaedia, ρ_γ goes as T^4. But this means in the case of radiation dominance that

$$\frac{\dot{T}(t)}{T(t)} = -AT^2(t) \tag{2.45}$$

where A is a positive constant whose value is of no concern to us for the moment. We can once again solve this equation by substituting $T(t) \sim t^n$ and solving for n. We leave it as an exercise to show that $n = -\frac{1}{2}$, i.e., that $T(t) \sim t^{-1/2}$ or that $R(t) \sim t^{1/2}$, which means that the universe was expanding more slowly during the radiation-dominated regime than during the matter-dominated regime.

We have stated several times that the universe is radiation-dominated for about 100,000 yr and then became matter-dominated. We now want to present the argument for this proposition. To a good approximation, the energy density of the matter we know about, at the temperatures that will be of interest to us, is determined by the number of protons. The question of the temperature is important. Recall from the Micropaedia (Eq. 1.8a) that the temperature equivalent to 1 GeV is about 10^{13} K. We will show that the crossover temperature to go from radiation to matter dominance is about 1 eV equivalent; i.e., about 10^4 K. This means that particles with less mass energy than 1 eV will, from the point of view of the energy density, be effectively massless at the time of this crossover. If the neutrinos, for example, have masses of this order, they will be included in that part of the energy density, along with the photons, that contributes to radiation dominance. We will make a careful examination of these questions in a later chapter. Here we want to make the crude assumption that only the photons contribute to the radiation density and only the protons contribute to the matter density. Since protons at a temperature of 1 eV are thoroughly nonrelativistic, their energy density is

$$\rho_M c^2 = m_p c^2 \times n_p \tag{2.46}$$

where n_p is the proton density. This quantity decreases as $1/R(t)^3$ or, in view of the discussion above, as T^3, while ρ_γ is decreasing as T^4. The fact that ρ_γ is decreasing with one power of T more than ρ_M is why there is a crossover of regimes. In view of the crudeness of our approximations we will take for ρ_γ the simple expression

$$\rho_\gamma \simeq kTn_\gamma \tag{2.47}$$

We have left out various numerical factors of order 1, and we leave it as an exercise for the reader to put these in and see how they affect the final answer. The crossover temperature T_c is determined by equating ρ_γ and ρ_M. With our approximations this produces the condition

$$kT_c \simeq m_p c^2 \times \frac{n_p}{n_\gamma} \tag{2.48}$$

But since both n_p and n_γ vary as T^3, the ratio is constant and can be evaluated for the present universe. From the discussion of Eqs. 1.44 and 1.46 of the Micropaedia we can write

$$kT_c \simeq \frac{3 \times 10^{-7}}{422} \times m_p c^2 \simeq 10^{-9} \times m_p c^2 \tag{2.49}$$

which implies that $kT_c \simeq 1$ eV, as advertised. This corresponds to a T_c of about 10^4 K. But how long did it take for the universe to cool off to this temperature? We will start counting from the Planck time, about 10^{-44} s, since before this time it is not certain whether space and time, as we understand them, are even defined. From the Planck time to the crossover time the universe was presumably radiation-dominated, which means that $T(t) \sim 1/t^{1/2}$. Thus

$$\frac{t}{t_P} = \left(\frac{T_P}{T}\right)^2 \tag{2.50}$$

or the crossover time t_c is given by

$$t_c = t_p \left(\frac{T_P}{T_c}\right)^2 \simeq 10^{12}\ \text{s} \simeq 10^5\ \text{yr} \tag{2.51}$$

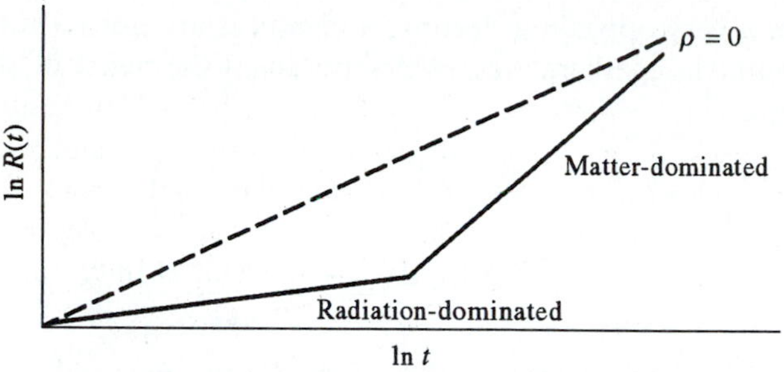

Figure 6 The development of $R(t)$ in the radiation- and then the matter-dominated regimes.

This is the origin of the claim that the regimes crossed over at about 100,000 yr. In Fig. 6, we give a log-log plot of the development of $R(t)$ in the two regimes, along with the unphysical case in which $\rho = 0$.

If we assume that the universe is matter-dominated for most of its lifetime we can, even in the cases in which E or k is not zero, draw some very interesting conclusions about the evolution of $R(t)$. Our starting point is a slight generalization of Eq. 2.28. For arbitrary t we see at once, for the matter-dominated regime and general Ω, that

$$H_0 t = \int_0^{\frac{R(t)}{R_0}} \frac{1}{[1 + \Omega(1/x - 1)]^{1/2}} dx \tag{2.52}$$

This integral is actually doable in closed form. Before undertaking this, let us rewrite Eq. 2.23 in the following way.

$$c^2 k = \dot{R}_0^2(\Omega - 1) \tag{2.53}$$

From this form of the equation the following correspondences are evident:

$$k = 1 \leftrightarrow \Omega > 1$$

$$k = 0 \leftrightarrow \Omega = 1$$

$$k = -1 \leftrightarrow \Omega = < 1$$

Exercising the "right of the professor" we shall work out the very interesting case with $\Omega > 1$ and leave the case with $\Omega < 1$ for the reader.

For $\Omega > 1$ the function $R(t)$ has a cycloidal behaviour (Fig. 7).

This means that the universe would undergo an endless cycle of Big Bangs and Big Crunches. One is reminded of the Buddhist notion of cycles of lives—the wheel

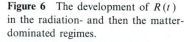

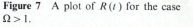

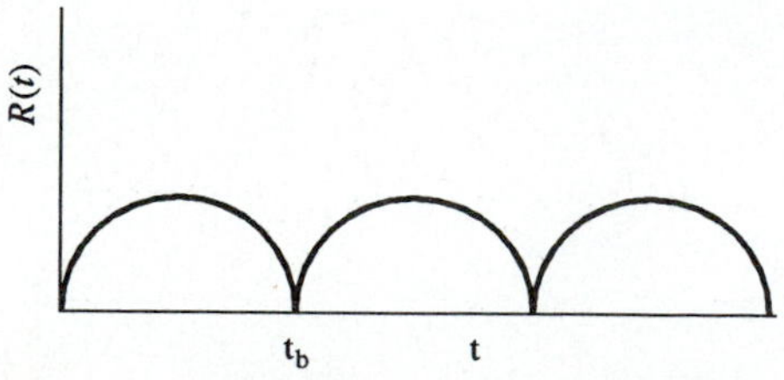

Figure 7 A plot of $R(t)$ for the case $\Omega > 1$.

of life. Some people find this a very appealing destiny, and others are not so sure. We shall content ourselves with the mathematics. We write down the result of the integration for one cycle. Thus

$$H_0 t = \frac{\Omega}{(\Omega-1)^{3/2}}\left(\frac{\pi}{2} - \tan^{-1}\left\{\left[1-\Omega\left(1-\frac{R_0}{R(t)}\right)\right]^{1/2} \times (\Omega-1)^{-1/2}\right\}\right) + \frac{R(t)}{R_0} \times \frac{[1-\Omega(1-R_0/R(t))]^{1/2}}{1-\Omega} \tag{2.54}$$

This somewhat forbidding-looking expression simplifies if we put in $t = t_0$ and solve for the present lifetime in terms of Ω and H_0. The following identity is also useful:

$$\frac{\pi}{2} - \tan^{-1}x = \cos^{-1}[x(1+x^2)^{-1/2}] \tag{2.55}$$

Thus

$$H_0 t_0 = \frac{\Omega}{(\Omega-1)^{3/2}} \times \cos^{-1}\Omega^{-1/2} + \frac{1}{1-\Omega} \tag{2.56}$$

This expression has two interesting limits. For $\Omega = 1$ we find at once that

$$H_0 t_0 = \tfrac{2}{3} \tag{2.57}$$

which is just Eq. 2.35, as one would expect. But for $\Omega \gg 1$ we have

$$H_0 t_0 \rightarrow \frac{\pi}{2\Omega^{1/2}} \tag{2.58}$$

This result is also plausible. It says that the more gravitating matter there is, the faster the universe runs through its cycle. We can also compute the time for the first bounce t_b by setting $R(t_b) = 0$ in Eq. 2.54. Thus

$$H_0 t_b = \frac{\pi\Omega}{(\Omega-1)^{3/2}} \tag{2.59}$$

For $\Omega \gg 1$ this becomes $\pi/\Omega^{1/2}$, which is just twice the value for $H_0 t_0$ given in the same limit in Eq. 2.58. We eagerly await the experimenters who will tell us what values of Ω and H_0 to put into these equations.

3

Ages

In this chapter we shall describe some of the methods used to assign ages to the various objects we observe. This discussion will bear some resemblance to our discussion of distances in that different techniques of increasing speculativeness are used as we go back further in time and as the objects under discussion are more and more remote from direct experimentation. Anything we can actually get our hands on, from meteorites to moon rocks, we can date pretty reliably. The method used is based on the fact that these objects contain, at least in part, long-lived radioactive isotopes. We will denote such a radioactive decay generically by $A \rightarrow B$, where A and B refer to the isotopes—mother and daughter—in question. Of course such decays will always produce, along with the daughters, some array of additional particles such as gamma rays, electrons, neutrinos, or alpha particles (helium nuclei). These will not concern us for the time being. Furthermore, we shall suppose that the daughter B is itself stable. The generalization to nonstable daughters is not difficult and can be found in standard nuclear physics textbooks. If we call $N_A(t)$ and $N_B(t)$ the number of A's and B's at any time t, then our concern will be to find these functions. Since we are assuming that there is only one daughter, we have necessarily the condition that

$$\frac{d}{dt}[N_A(t) + N_B(t)] = 0 \tag{3.1}$$

i.e., the total number of particles is conserved. Thus

$$N_A(t) + N_B(t) = N_A(0) + N_B(0) \tag{3.2}$$

One of the principal complications of radioactive dating is that we do not, as a rule, know $N_B(0)$. We must either make a plausible guess as to what it is or find a way around our lack of knowledge. For $N_A(t)$ we can assume the exponential decay law

$$N_A(t) = N_A(0)e^{-\lambda_A t} \tag{3.3}$$

where the decay rate λ_A is related to the lifetime τ_A by

$$\lambda_A = \frac{1}{\tau_A} \tag{3.4}$$

Using Eq. 3.2, we can write down at once the fundamental law of radioactive dating:

$$N_B(t) = N_A(t)(e^{\lambda_A t} - 1) + N_B(0) \tag{3.5}$$

To see how it works, let us first find the age of the earth. In this case what is used is a long decay chain beginning with the uranium isotope ^{238}U and ending with the stable lead isotope ^{206}Pb. In the course of this series of decays, eight alpha particles are produced and remain trapped in the rock. Most of the decays in the chain are short-lived. The exception is the decay $^{238}U \rightarrow {}^{234}Th + \alpha$. It has a lifetime of 4.5×10^9 yr, and it is this lifetime that determines the rate of the buildup of the helium. Hence we write

$$N(t)_{He} \simeq N_{^{238}U}(t)(e^{\lambda_{^{238}U} t} - 1) \tag{3.6}$$

The t that enters here is the time interval between the formation of the uranium in the earth's crust and the present, and this is the time we want to determine. But we can measure the amount of both helium and uranium presently in the rock formation. Since we know $\lambda_{^{238}U}$ from laboratory measurements we can solve Eq. 3.6 for t. We have made the assumption that there was essentially no helium present in the earth rocks at $t = 0$. In some cases one can relax the assumption that none of the daughter elements existed initially. This complicates the analysis, and the interested reader should consult one of the standard texts on nuclear physics to see how it is done. The people who do this work report that $t \simeq 3.7 \times 10^9$ yr. This is a measure of the age of the earth, and the earth must be at least as old as this.

Radioactive dating is also used to date archaeological artifacts such as mummies. What is used here is a radioactive isotope of carbon, ^{14}C, with a half-life of 5730 yr. This isotope is being produced continuously as the earth is bombarded by cosmic rays. The isotope ^{14}C is absorbed as part of the metabolic processes of living systems. When such a system dies, it no longer absorbs the isotope, and its activity $A(t)$, defined in terms of the number of nuclei present at time t, $N(t)$, and the decay rate λ by the equation

$$A(t) \doteq \lambda N(t) = \lambda N(0)e^{-\lambda t} \tag{3.7}$$

continues to decrease with the passage of time. If we assume that $N(0)$ for ^{14}C is about the same as it is in the present atmosphere—it is being produced by the same mechanisms—we can again solve for t. This method works for objects not much older than 30,000 yr. After that, the activity becomes too small to be measurable.

Next we turn to the dating of meteorites. Here we make use of three facts:

a. ^{235}U eventually decays into ^{207}Pb.
b. ^{238}U eventually decays into ^{206}Pb.
c. No long-lived decay process produces ^{204}Pb. What you see now is, more or less, what was produced originally.

The idea then is to form two ratios R_1 and R_2 defined as follows:

$$R_1(t) = \frac{N(t)_{\{^{207}Pb\}}}{N(t)_{\{^{204}Pb\}}} = \frac{N(0)_{\{^{207}Pb\}}}{N(t)_{\{^{204}Pb\}}} + \frac{N(t)_{\{^{235}U\}}}{N(t)_{\{^{204}Pb\}}}(e^{\lambda_{\{^{235}U\}} t} - 1) \tag{3.8}$$

and

$$R_2(t) = \frac{N(t)_{\{206\text{Pb}\}}}{N(t)_{\{204\text{Pb}\}}} = \frac{N(0)_{\{206\text{Pb}\}}}{N(t)_{\{204\text{Pb}\}}} + \frac{N(t)_{\{238\text{U}\}}}{N(t)_{\{204\text{Pb}\}}}(e^{\lambda\{238\text{U}\}t} - 1) \tag{3.9}$$

The ratios R_1 and R_2 can be measured in two different meteorites which presumably were formed at about the same time—around the time of formation of the Solar System. (An attractive speculation is that meteorites are fragments of asteroids and that these in turn are pieces of a planet that exploded early in the history of the Solar System.) Let us call these meteorites a and b. Since they have, it is assumed, the same initial lead compositions, if we take the difference of these ratios between the two meteorites, the references to the quantities involving lead at the time of meteorite formation, cancel out. We are left with

$$\frac{R_{1a}(t) - R_{1b}(t)}{R(t)_{2a}(t) - R_{2b}(t)} = \frac{e^{\lambda\{235\text{U}\}t} - 1}{e^{\lambda\{238\text{U}\}t} - 1} \times \frac{\dfrac{N_a(t)_{\{235\text{U}\}}}{N_a(t)_{\{204\text{Pb}\}}} - \dfrac{N_b(t)_{\{235\text{U}\}}}{N_b(t)_{\{238\text{U}\}}}}{\dfrac{N_a(t)_{\{238\text{U}\}}}{N_a(t)_{\{204\text{Pb}\}}} - \dfrac{N_b(t)_{\{238\text{U}\}}}{N_b(t)_{\{204\text{Pb}\}}}} \tag{3.10}$$

All the quantities in this somewhat lengthy formula are measurable and, experimenters tell us, produce an age for the meteorites of 4.55×10^9 yr. This is greater than the rock crust age given above. That is sensible because, as far as we know, the meteorites were formed at about the same time as the planets, while it took some additional time for the earth rocks to cool down and segregate out. By the way, the moon rocks are about as old as the meteorites.

We turn next to the ages of stars and globular clusters of stars. To deal with this we must give something of the life history of a star. For this purpose we once again draw the Hertzsprung-Russell diagram, but this time in somewhat greater detail (Fig. 1).

Stars begin their lives as contracting gasses of light nuclei. As such a gas cloud contracts, it loses gravitational potential energy. To conserve the total energy, this energy loss is made up by the increasing kinetic energy of the particles making up the gas. In other words, the gas gets heated up. Working out the complicated details of this contraction process occupies the time of specialists, but this is the general idea. If the cloud has a mass of greater than about 0.08 $M_\odot$, then the temperature can be raised to the point at which nuclear fusion reactions of the type that we described in the helium production section of the Micropaedia begin to take hold. This critical temperature can be as low as 1-keV equivalent. At first sight, it is puzzling that this temperature is as low as it is. The reason why we might be surprised is that for a fusion reaction to take place, the nuclei in question must come within the range of the attractive nuclear force R, about 10^{-13} cm. But these nuclei also repel each other electrostatically. The Coulomb energy of this repulsion is of order $e^2/R \simeq 1$ MeV. But, from the Micropaedia (Eq. 1.8a), we see that 1 MeV corresponds to a temperature of about 10^{10} K—much hotter than the actual temperatures at which fusion begins. One of the reasons for this discrepancy lies in the inadequacy of the classical calculation. In quantum mechanics, particles can tunnel through the

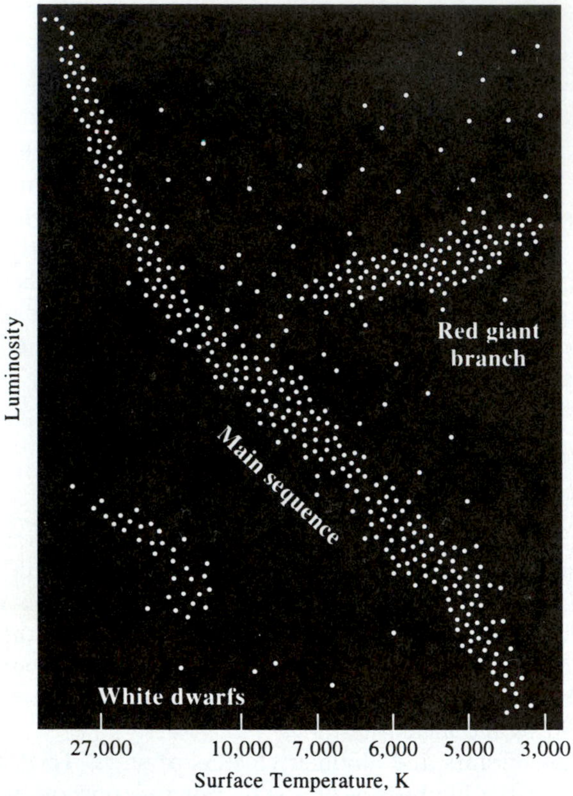

Figure 1 The Hertzsprung-Russell diagram revisited.

Coulomb barrier—barrier penetration—and this requires less energy than the classical calculation would predict. Some of the reactions that can occur have been discussed in the Micropaedia. But we can also have reactions involving, for example, lithium and berylium such as

$$^{7}\text{Li} + p \rightarrow {}^{4}\text{He} + {}^{4}\text{He}$$

or

$$^{9}\text{Be} + p + p \rightarrow {}^{3}\text{He} + {}^{4}\text{He} + {}^{4}\text{He}$$

The latter reaction is interesting because it involves a collision among three bodies in the initial state. Such collisions are next to impossible to arrange in a terrestrial physics laboratory, but they play an important role in the life cycle of stars because of the extreme conditions of temperature and density in the interior of stars, to say nothing of the fact that there are eons of time available for these reactions to take place. At about 10 million kelvins, the star becomes hot enough so that the proton-proton fusion cycle, leading to the stellar helium production described in the Micropaedia begins. The relatively massive helium that is produced sinks down into the central portion of the star (Fig. 2).

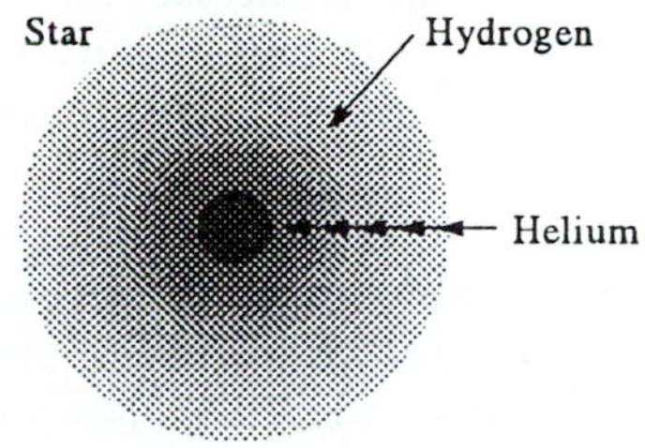

Figure 2 A typical main-sequence star with its helium core.

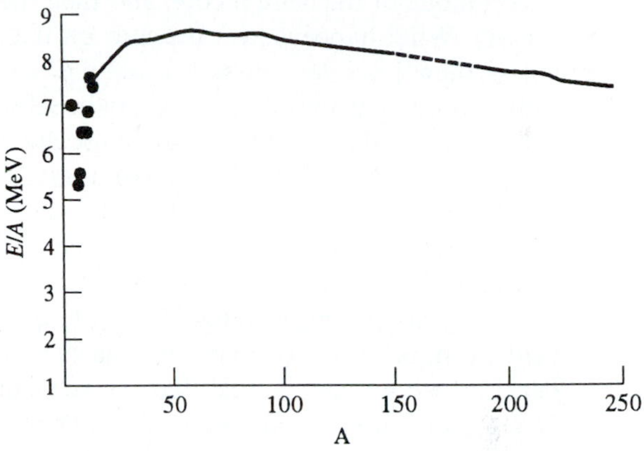

Figure 3 The curve of binding energy.

The star then begins to move along the main sequence of the Hertzsprung-Russell diagram. What happens next depends on the mass of the star. If it is greater than about one solar mass, when about 10 percent of the hydrogen has been turned into helium, the core becomes inert from the point of view of fusion reactions. But its temperature continues to increase as it contracts gravitationally. Densities can reach as much as $10^5\ \mathrm{g\,cm^{-3}}$. The energy production then takes place in the outer part of the star, and it begins to swell. Therefore its luminosity increases owing to its enlarged surface area, even though its temperature remains constant. It becomes what is known as a red giant, and it moves off the main sequence of the Hertzsprung-Russell diagram. Meanwhile the core continues to contract and heat up. When it reaches a temperature of about 100 million degrees the helium can itself begin to fuse. One has reactions like

$$^{4}\mathrm{He} + {}^{4}\mathrm{He} \rightarrow {}^{8}\mathrm{Be} + \gamma$$

or

$$^{8}\mathrm{Be} + {}^{4}\mathrm{He} \rightarrow {}^{12}\mathrm{C} + \gamma$$

Eventually elements like ^{16}O, ^{20}Ne, and ^{24}Mg are produced. But like all good things, this nucleus building by successive fusion reactions comes to an end, in this case in the neighborhood of iron. This has to do with the so-called curve of binding energy. If we plot the binding energy per nucleon as a function of the atomic number A of successive nuclei, the curve rapidly rises for the low-mass nuclei to about 8.7 MeV per nucleon and then begins to flatten out before it turns over and starts to fall gently at $A = 50$ (Fig. 3).

This means that it is energetically advantageous for nuclei lighter than iron to fuse with other light nuclei to make still more tightly bound nuclei and to release the mass-energy difference in the form of particles like gamma rays or neutrinos. On the other hand, for heavy nuclei, it is energetically profitable to split into lighter, more tightly bound nuclei—hence nuclear fission. But, returning to the stars, the nuclear fusion mode of the helium core, and then the other light elements, is over with quite rapidly. What happens next depends on the mass of the star. If the mass of the star is less than 1.4 solar masses—a number called the Chandrasekhar limit (after the Indian-born astrophysicist S. Chandrasekhar who first derived it), the star can collapse into what is known as a white dwarf. The radii of these stars is a function of their mass. For white dwarfs with masses comparable to that of the sun, the radii are about 100 times smaller than the sun's radius. Some white dwarfs have core densities 100,000 times greater than that of the sun. These stars no longer contract but simply go on cooling indefinitely.

Stars in the mass range $1.2 \leqslant M/M_{\odot} \leqslant 2$ collapse into what are known as neutron stars. These are amazing objects that resemble gargantuan nuclei. They have densities about 10^{14} times that of the sun but have diameters of only about 30 km—sort of like Manhattan. But they rotate with periods as small as milliseconds with enormous regularity. It would take us too far afield from our cosmological purposes to describe these wonderful objects further. The interested reader can find a great deal of literature at all levels about them. Likewise, we cannot enter very deeply into the subject of black holes. Theory predicts that if the mass of a star satisfies the condition $M/M_{\odot} > 3$, then nothing can stop the gravitational collapse of the core. The object that is produced is so gravitationally dense that, at least, according to Einstein's classical theory of gravitation, no light can escape from it. The quantum theory predicts that these objects can radiate, but that again would take us too far afield to go into. Many astronomers are persuaded that there is evidence for the existence of black holes, and others remain skeptical. Black holes would certainly enter our cosmological story in a big way if it turned out that there were so many of them that they provided the missing dark mass to allow $\Omega = 1$; i.e., to close the universe. We must wait and see.

Now we return to out star dating. We can estimate low long a star remains on the main sequence in the following way. The binding energy of helium is 28.28 MeV, and the proton rest mass energy is 938.26 MeV, to sufficient accuracy for our purposes. We will also ignore the neutron-proton mass difference. Thus the energy produced per gram of helium from neutrons and protons is approximately 6.8×10^{18} erg g^{-1}—using the conversion factors from the Micropaedia. Let us call the initial mass of the star M and let us suppose it emits L ergs per second of radiation as a result of these fusion mechanisms. Astronomers tell us that, after about 12 percent of this mass is lost, the star begins to move off the main sequence and enters the red giant phase. Thus we can write, using overall energy conservation,

$$6.8 \times 10^{18} \times 0.12M = Lt_s \tag{3.11}$$

where t_s is the time at which the star leaves the main sequence. This time gives us a sort of scale for star ages. We can rewrite the equation for t_s (in years) in the following way:

$$t_s = 0.26 \times 10^{11} \text{ yr} \times \frac{M/M_{\odot}}{L/L_{\odot}} \times \frac{\mathrm{M}_{\odot}}{L_{\odot}} \tag{3.12}$$

where, to normalize things, we have introduced the sun's mass $M_\odot \simeq 2 \times 10^{33}$ g and the sun's luminosity $L_\odot \simeq 4 \times 10^{33}$ erg s^{-1}. Thus

$$t_s \simeq 1.3 \times 10^{10}\ \text{yr} \times \frac{M/M_\odot}{L/L_\odot} \tag{3.13}$$

If we recall from the previous chapter (Eq. 2.46) that for a matter-dominated universe, with $\Omega = 1$, the present lifetime of the universe is given by

$$t_0 = 6.13 \times 10^9\ \text{yr h}^{-1} \tag{3.14}$$

we see that these stellar ages appear to favor the smaller values of h. We next turn to the detailed working-out of various cosmological processes as the universe expands and ages.

4

Equilibrium

We are now prepared to study the evolution of a variety of the specific processes that take place in the early universe. One of the great simplifications that will help us in this work is that, at least for certain time periods, these processes occur under equilibrium conditions. For example, as we shall see, until the universe has cooled down to a temperature of about 10^4 K, the electrons and photons are in a state of thermal equilibrium. The photons act as if they are trapped inside the walls of a blackbody. The details of the structure of this blackbody don't matter. So long as there is one, its temperature is enough to specify its physics. The question we want to address is, How do we know under which circumstances such a blackbody exists or, more generally, how can we determine when processes are in equilibrium in the early universe? Generally speaking, these equilibrium conditions are set up when the particles in question undergo many collisions in a short time. But how many collisions? And in how short a time? One possible criterion to adopt is that, as a particle traverses the universe, it has to undergo at least one collision. For the case of the open universe ($\Omega < 1$) the actual size of the universe is infinite. So, with this in mind, we define the size of the universe, as far as determining equilibrium is concerned, as the size of the visible universe. Since we will not be concerned with factors of order 1—we will only be making order-of-magnitude estimates—we will take for this size d, where

$$d = c\frac{R(t)}{\dot{R}(t)} \tag{4.1}$$

In the early universe, when the temperature is very high, most of the particles we will be interested in will be effectively massless. By this we mean that for such particles, $kT/mc^2 \gg 1$. But kT is related to the average speed of these particles. Recall that for a classical gas in equilibrium, its average kinetic energy is proportional to its temperature. Thus the objects we will be interested in are moving, on the average, at speeds close to that of light. Therefore, from Eq. 4.1, we see that the time it takes for such a particle to cross the visible universe is about $R(t)/\dot{R}(t)$. Now, let us call the rate at which some given reaction takes place ω. Then our criterion for equilibrium boils down to the condition that

$$\frac{R(t)}{\dot{R}(t)} \simeq \frac{1}{\omega} \tag{4.2}$$

or

$$\omega \simeq \frac{\dot{R}(t)}{R(t)} \tag{4.3}$$

which guarantees that there will be at least one interaction each time the particle traverses the visible universe. As we shall see, the rate ω depends on time because it depends on the temperature of the reacting particles, and this temperature changes in time as the universe expands. That there should be such a temperature dependence of ω is plausible. Keep in mind that the average energies of the particles that are reacting depend on the temperature, and that, in general, the reaction rates depend on the energies of these particles, hence on their temperature. In Chap. 2 we described how one goes about estimating $\dot{R}(t)/R(t)$. Here we want to describe how one finds ω. To be as specific as possible, let us study the following idealized situation. We imagine a cylinder of circular area A filled with particles with a number density n (the number of particles per unit volume in the cylinder), all of them moving in the same direction with the speed of light. Of course, in a more realistic situation, the particles will move with various speeds in various directions. One will have to carry out an averaging procedure, which will change our results by numerical factors of order 1. But we are ignoring such factors in these order-of-magnitude estimates. We may ask, in a time interval Δt, how many of these particles will cross the area A. Figure 1 illustrates the situation.

Under the conditions we have specified, all the particles in the cylinder we have drawn will pass through the area A in the time interval Δt. In other words, the number of particles that will pass through A is $n\,\Delta t c A$. This means that the number that will pass through, per unit area per second, which we call the flux $\mathscr{F}$, is given by

$$\mathscr{F} = nc \tag{4.4}$$

More generally, if $\langle|\mathbf{v}|\rangle$ is the average speed of the particles in question, the flux is defined as

$$\mathscr{F} = n\langle|\mathbf{v}|\rangle \tag{4.5}$$

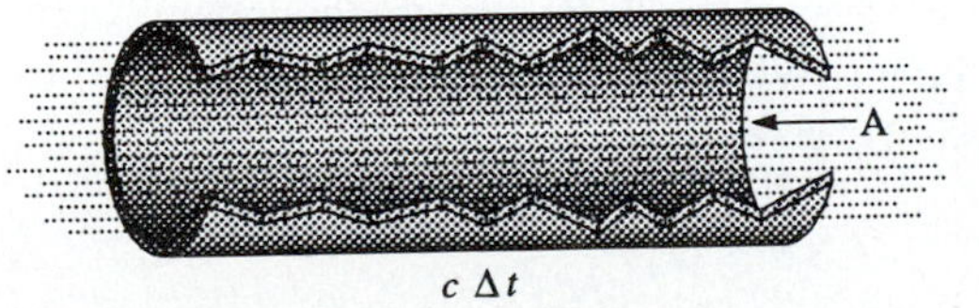

Figure 1 A cylindrical flux tube containing particles all moving in the same direction with speed c.

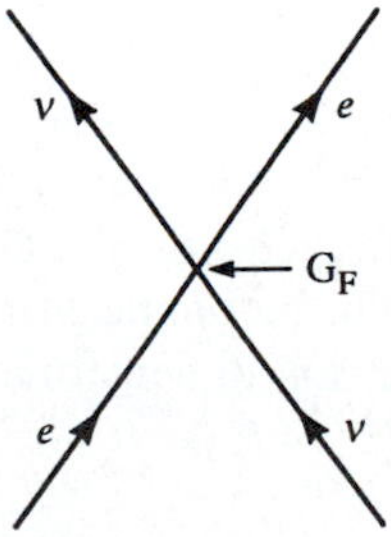

Figure 2 The point contact interaction between electrons and neutrinos.

where we have introduced the notation $|\mathbf{A}|$ for the length of any vector $\mathbf{A}$, and the symbol $\langle\ \rangle$ for some averaging procedure.

The reaction rate ω must be proportional to $\mathscr{F}$. To take an extreme case, suppose that there was no flux of particles impinging on the target. Then, obviously, there could be no reactions. The proportionality factor between ω and $\mathscr{F}$ has the dimensions of an area and is conventionally called σ, the cross section. While the cross section has the dimensions of an area, it is not really related to the geometrical area of the target particles. The following is a well-known example: if one scatters (collides) nonrelativistic quantum mechanical particles from an impenetrable hard sphere of radius R, then at low energies, where the de Broglie wavelength of these particles is much larger than R, the cross section turns out to be $4\pi R^2$, which is the geometrical surface area of the sphere. But at high energies, where the wavelength is much less than R, the cross section becomes $2\pi R^2$, which can be understood only by working out the quantum mechanical theory of the scattering. In general, σ depends on the forces that are causing the interactions. Indeed, measuring σ is often the best, if not the only, way of learning about these forces. To compute σ from first principles is properly the subject matter of a quantum mechanics or, in the relativistic context, a quantum theory of fields, course. We shall in the spirit of this book, however, give simple dimensional arguments that will lead us to the order-of-magnitude results we are interested in. To illustrate our methodology, let us pose the following question. We know that in the early universe there were neutrinos and electrons present in about the same abundance as the photons that eventually become the presently observed 3 K radiation. As discussed in the Micropaedia, these neutrinos and electrons interact through weak interactions. If the temperatures are such that the average kinetic energies of the neutrinos and electrons are much less than the mass energies of the intermediate weak bosons, say 80 GeV—which corresponds to a temperature of about 10^{15} K, then the point contact Fermi theory described in the Micropaedia works perfectly well (Fig. 2).

The coupling shown in Fig. 2 depends on the Fermi constant G_F to the first power. But this diagram represents the probability amplitude. The probability is obtained by squaring this amplitude. This means that for this process the rate ω, hence the cross section σ, depends on the square of G_F. But we saw in the Micropaedia (Eq. 1.36) that G_F has the dimensions of energy $\times$ centimeter^3. We may then assign ourselves the following task of dimensional analysis. We want to construct a quantity with the dimensions of an area, σ, out of G_F^2, $\hbar$, c, and kT. The latter gets into the act because we expect σ to be energy-dependent, which means that it will be temperature-dependent. We leave it as an exercise for the reader to carry out this dimensional analysis. The result is

$$\sigma \simeq G_F^2 \frac{(kT)^2}{(\hbar c)^4} \tag{4.6}$$

We use the $\simeq$ symbol because the dimensional analysis does not fix the numerical factors, which are of order 1, but only the dimensional factors. Next we determine the flux. In the temperature regime of interest, the particles will be effectively massless in the sense discussed earlier. Hence their number density n will have the same dependence on the temperature T as the photon number density; i.e.,

$$n \simeq \left(\frac{kT}{\hbar c}\right)^3 \tag{4.7}$$

This result is also a consequence of dimensional analysis. This quantity has the dimensions of an inverse volume. The average speed of these particles is c, so that their flux $\mathcal{F}$ is given by

$$\mathcal{F} \simeq \left(\frac{kT}{\hbar c}\right)^3 c \tag{4.8}$$

If we combine Eq. 4.8 and 4.6, we have

$$\omega \simeq G_F^2\left(\frac{kT}{\hbar c}\right)^5 \times \frac{c}{(\hbar c)^2} \tag{4.9}$$

We can use the discussion associated with Eq. 1.37 of the Micropaedia to write Eq. 4.9 in the more useful form

$$\omega \simeq \left(\frac{kT}{m_p c^2}\right)^5 \frac{m_p c^2}{\hbar} \times 10^{-10} \tag{4.10}$$

where we have introduced the mass of the proton m_p as a convenient scale.

We want to equate this expression for ω to $\dot{R}(t)/R(t)$. We will be working in the radiation-dominated regime in the sense described in Chap. 2 of the Macropaedia. In this regime the energy density ρ is dominated by effectively massless particles. So, by dimensional analysis,

$$\rho \simeq kT\left(\frac{kT}{\hbar c}\right)^3 \tag{4.11}$$

which is the average energy multiplied by the average number of particles per unit volume—always of course up to numerical factors of order 1. But the Einstein–Friedmann equation (see again Chap. 2 of the Macropaedia) tells us, up to numerical factors, that

$$\left[\frac{\dot{R}(t)}{R(t)}\right]^2 \simeq \rho \times \frac{\hbar c}{M_P^2 c^2} \tag{4.12}$$

where M_P is the Planck mass. Thus, using Eq. 4.11, we have

$$\frac{\dot{R}(t)}{R(t)} \simeq \frac{(kT)^2}{M_P} \frac{1}{\hbar c^2} \tag{4.13}$$

Using Eq. 1.11 of the Micropaedia, we can write this as

$$\frac{\dot{R}(t)}{R(t)} \simeq 10^{-19}\left(\frac{kT}{m_p c^2}\right)^2 \frac{m_p c^2}{\hbar} \simeq \left(\frac{kT}{m_p c^2}\right)^2 10^5\ \text{s}^{-1} \tag{4.14}$$

If we equate 4.10 and 4.14, we have

$$\left(\frac{kT_c}{m_p c^2}\right)^3 \simeq 10^{-9} \tag{4.15}$$

where T_c is the critical temperature at which the neutrino-electron interactions fall out of equilibrium. Above this temperature $\dot{R}/R$ was greater than $n\sigma c$, so that there were many electron-neutrino collisions as either the electron or the neutrino crossed the universe. These many collisions are what produce equilibrium. That there is a critical temperature below which the particles are no longer in equilibrium can be traced to the fact that $\dot{R}(t)/R(t)$ falls off as T^2 in the radiation-dominated regime, while the weak interaction rate falls off as T^5 in the same regime. Thus $kT_c \simeq 10^{-3} m_p c^2$, which corresponds to a temperature of about 10^{10} K. We may use Eq. 4.14 to find the approximate time at which this critical temperature is realized. Thus

$$t_c \simeq \frac{R(t)}{\dot{R}(t)} \simeq 10^{-5}\left(\frac{m_p c^2}{kT_c}\right)^2 s \simeq 10\ \text{s} \tag{4.16}$$

When we come to helium production in Chap. 6 we will construct a somewhat more accurate clock which reduces this time to somewhat less than a second. This means that, until about a second after the Big Bang, the electrons and neutrinos are in equilibrium.

Whenever effectively massless particles like the neutrinos and electrons are in equilibrium, we can make some very general statements about them. We shall use some elementary results from statistical mechanics. We refer the reader to any of the standard texts. Let us call the number of massless particles of some specific type, per unit volume—with a momentum in the interval between $\mathbf{p}$ and $\mathbf{p} + d\mathbf{p}$, $n(\mathbf{p})$. Then the number density of particles, n, is related to $n(\mathbf{p})$ by the equation

$$n = \int \frac{d^3p}{(2\pi\hbar)^3} n(p) \tag{4.17}$$

We have dropped the vector notation since in a homogeneous, isotropic universe n is only a function of the magnitude of $\mathbf{p}$. The energy density of these particles, ρ, is given by

$$\rho = \int \frac{d^3p}{(2\pi\hbar)^3} pcn(p) \tag{4.18}$$

Note that for such a massless particle, $\langle|\mathbf{p}|\rangle$, the average value of the magnitude of the momentum, obeys the relation

$$\langle|\mathbf{p}|\rangle = \frac{\rho}{c} \tag{4.19}$$

We shall use this result to find the pressure exerted by these massless particles. Suppose that, at the end of the flux tube shown in Fig. 1, we actually put a barrier that covers that end. What will be the force per unit area on such a barrier? We will consider the somewhat more realistic situation in which the particles can move in any direction within the tube. If they were all moving in the same direction, then, as we have seen, in a time interval Δt a number equal to $nc\ \Delta t A$ would hit the barrier, where n is the number density of particles in the tube. In the more realistic situation we are now contemplating, on the average only one-third of the particles are moving along the axis of the cylinder and, of these, only one-half are moving toward the barrier. Thus the number that would hit the barrier is only $\frac{1}{6}nc\ \Delta t\ A$. If we assume that the collisions with the barrier are strictly elastic then, on the average, each collision would transfer a momentum of $2\langle|\mathbf{p}|\rangle$ to the barrier. But, from Eq. 4.19, this is $2\rho/c$. To compute the pressure we recall that it is the momentum transferred per second per unit area to the barrier. Thus we arrive at the famous result that the pressure P on the barrier due to these massless particles is given by

$$P = \frac{\rho}{3} \tag{4.20}$$

This means that at high temperatures, where the particles are relativistic, the pressures and the energy densities are comparable.

The situation is very different at low temperatures, $kT \ll mc^2$, where the particles are nonrelativistic. Once again, the number of particles that strike the end of the tube in the interval Δt is given in terms of the average speed by $\frac{1}{6}\langle|\mathbf{v}|\rangle\ \Delta t\, nA$. On the average, these particles transfer a momentum of $2m\langle|\mathbf{v}|\rangle$. Thus the pressure P is given by

$$P = \tfrac{1}{3}nm\langle|\mathbf{v}|\rangle^2 \simeq \tfrac{2}{3}n \times \tfrac{1}{2}m\langle\mathbf{v}^2\rangle \tag{4.21}$$

We have used the $\simeq$ symbol in the last equation because $\langle|\mathbf{v}|\rangle^2$ and $\langle\mathbf{v}^2\rangle$ differ by factors of order of magnitude 1. A more rigorous calculation actually produces the $\langle\mathbf{v}^2\rangle$ result. Again, the reader is referred to a standard statistical mechanics text. But $m/2\langle\mathbf{v}^2\rangle$ is the average kinetic energy. For a perfect gas in three dimensions this is equal to $3/2nkT$. Thus we have, as we would expect, the equation of state for a perfect gas

$$P = nkT \tag{4.22}$$

Let us compare P to the rest energy density nmc^2. Clearly $P/nmc^2 \simeq kT/mc^2 \ll 1$. Hence we see that at low temperatures, where "low" is defined in terms of the temperatures and rest masses involved, we can neglect the pressure of the particles.

The place where we have to be careful about taking the pressure into account is in computation of the entropy. We begin this calculation by examining the conservation of energy, i.e., the first law of thermodynamics. During most of the universe's expansion we do not expect the entropy to change. We do, however, expect that at certain selected epochs there will be very large amounts of entropy generated in very short time intervals—probably by the creation of particles—for otherwise how can we account for the fact that there is such a huge entropy within the visible universe (about 10^{71} erg K^{-1}) at the present time? (If we divide this dimensional number by Boltzmann's constant, we have the even huger dimensionless number 10^{87}.) These entropy-generating epochs, we will argue in due course, must have occurred in the very early universe, otherwise we would have difficulty understanding why the cosmic background radiation is so blackbody-like. These are ideas that we will come back to, but for present purposes, let us suppose that no "friction" is being generated as the universe expands. Hence the energy conservation equation reads

$$dE + P\,dV = 0 \tag{4.23}$$

where E is the *total* energy, P is the pressure, and dV is the change in volume. In the context of the expanding universe we can write this as

$$\frac{d}{dt}[R(t)^3\rho] + P\frac{d}{dt}[R(t)^3] = 0 \tag{4.24}$$

This is often rewritten as

$$\frac{d}{dt}[R(t)^3(P+\rho)] = R(t)^3\frac{d}{dt}P \tag{4.25}$$

The time dependence of the pressure and energy density come in implicitly because of their dependence on the temperature. We can study this equation in two interesting limits:

a. The "low"-temperature limit in which $P \simeq 0$. Thus Eq. 4.25 becomes

$$\frac{d}{dt}[R(t)^3\rho] = 0 \tag{4.26}$$

This result is entirely consistent with the universe's being matter-dominated, which led us to neglect the pressure in the first place. In this universe, $\rho \simeq nmc^2$, where m is, for example, the proton mass. Thus Eq. 4.26 is simply a restatement of particle conservation.

b. The "high" temperature limit in which $P = P/3\rho$. In this limit Eq. 4.24 becomes

$$\frac{d}{dt}[R(t)^3\rho] + \frac{\rho}{3}\times\frac{d}{dt}[R(t)^3] = 0 \tag{4.27}$$

or

$$\frac{\dot{R}(t)}{R(t)} = -\tfrac{1}{4}\times\frac{\dot{\rho}(t)}{\rho(t)} \tag{4.28}$$

Thus, in this limit, $\rho(t) \sim 1/R(t)^4$. This is consistent with the proposition that $\rho(t) \sim T(t)^4$ provided that $T(t) \sim 1/R(t)$, a result that we are familiar with. See the discussion relating to Eq. 2.53.

We now turn to the question of entropy. These thermodynamic arguments are always confusing because one often finds oneself wondering what is a function of what. Let us for the moment ignore the context of the expanding universe and simply consider a gas confined to a volume V at temperature T. All the other thermodynamic quantities will be functions of V and T. But we will specialize to systems in which the energy density ρ is a function of T alone. For example, for massless particles in equilibrium, $\rho \sim T^4$ and the volume does not enter the expression for the energy density at all. For these systems, and this is what we are going to prove, the entropy density s takes a very simple form, namely

$$s = \frac{P + \rho}{T} \tag{4.29}$$

which enables us to compute the entropy of such systems readily. We begin our discussion with a general equation for the change in the total entropy, $S = Vs$, where $S = S(V, T)$. This equation is the generalization of Eq. 4.23, where now we include the heat energy due to "friction". Thus

$$dS(V, T) = \frac{1}{T}[d(\rho(V, T)V) + P(V, T)dV] \tag{4.30}$$

This equation is perfectly general. But now we use the assumption that $\rho(V, T) = \rho(T)$. Hence

$$\begin{aligned} dS(V, T) &= \frac{1}{T}\left\{V\frac{\partial\rho(T)}{\partial T}dT + [\rho(T) + P(V, T)]\,dV\right\} \\ &= \frac{\partial S(V, T)}{\partial V}dV + \frac{\partial S(V, T)}{\partial T} \times dT \end{aligned} \tag{4.31}$$

The second equation simply expresses the total differential of $S(V, T)$. We now invoke the condition that

$$\frac{\partial^2}{\partial V\,\partial T}[S(V, T)] = \frac{\partial^2}{\partial T\,\partial V}[S(V, T)] \tag{4.32}$$

along with the statements that we can read off from Eq. 4.31:

$$\frac{\partial}{\partial T}S(V, T) = \frac{V}{T}\frac{\partial}{\partial T}\rho(T) \tag{4.33a}$$

and

$$\frac{\partial}{\partial V}S(V, T) = \rho(T) + P(V, T) \tag{4.33b}$$

Thus

$$\frac{\partial^2}{\partial V\,\partial T}[S] = \frac{\partial\rho}{\partial T}\frac{1}{T} = \frac{\partial^2}{\partial T\,\partial V}[S] = \frac{\partial}{\partial T}\left[\frac{1}{T}(\rho + P)\right] \tag{4.34}$$

or

$$\frac{\partial}{\partial T}P = \frac{1}{T}(P + \rho) \tag{4.35}$$

Hence we can now evaluate $d(PV)$:

$$d(PV) = dV\frac{\partial}{\partial V}(PV) + dT \times \frac{\partial}{\partial T}(PV) = dVP + \frac{dT}{T}(P + \rho)V \tag{4.36}$$

Referring back to Eq. 4.30, we see that, with our assumptions, the entropy density s obeys the equation

$$ds = d\left(\frac{\rho + P}{T}\right) \tag{4.37}$$

This determines s up to an additive constant:

$$s = \frac{\rho + P}{T} + \text{constant} \tag{4.38}$$

Equation 4.38 fixes the entropy differences. Since the constant drops out of these differences, we will drop it in our expressions for the entropy. In particular, using Eq. 4.20, we have for massless particles,

$$s = \frac{4}{3} \times \frac{\rho}{T} \tag{4.39}$$

We will make good use of this result shortly.

To conclude this discussion we will demonstrate the consistency between Eq. 4.37 and the conservation equation (Eq. 4.25); that is,

$$\frac{\partial}{\partial t}[R^3(P + \rho)] = R^3 \frac{\partial}{\partial t} P = R^3 \frac{\partial T}{\partial t}\frac{\partial P}{\partial T} \tag{4.25}$$

Using Eq. 4.35, we can write this as

$$\frac{\partial}{\partial t}[R^3(P + \rho)] = R^3 \frac{\partial T}{\partial t}\frac{1}{T}[R^3(P + \rho)] \tag{4.40}$$

But this is simply the statement that

$$\frac{\partial}{\partial t}\left[\frac{R^3(P + \rho)}{T}\right] = 0 \tag{4.41}$$

In order words, when no heat energy is generated as the universe expands, entropy is conserved.

Up to now we have been content to make order-of-magnitude arguments involving dimensional analysis. We would now like to make some of this discussion more precise. Here again, we shall be using results that are derived in standard statistical mechanics courses. One of the great discoveries of the 1920s was that of quantum statistics. It turns out that all elementary particles fall into one of two statistical groups—Fermi-Dirac or Bose-Einstein. Every particle has a "spin", an angular momentum that persists even when the particle is brought to rest. Spins are conveniently measured in units of $\hbar$. Bose-Einstein particles have even-integer—always in units of $\hbar$—spins. For example, the photon has spin 1, while the pi meson has spin 0. On the other hand, Fermi-Dirac particles have half odd-integer spins. For example, the electron, the neutrino, the neutron, and the proton all have spin $\frac{1}{2}$. Remarkably, there is a deep connection between the spin of a particle and its statistics. Broadly speaking, Bose-Einstein particles can assemble together in a single quantum state. The most striking example of this is liquid helium which at temperatures below a critical temperature of 2.17 K condenses into a state that can exhibit frictionless collective motion, superfluidity. On the other hand, Fermi-Dirac particles can never be in a state where any two of them have exactly the same quantum numbers—something that is known as the Pauli exclusion principle. Using the exclusion principle one can give an accounting of the periodic table of elements, for example. These statistical distinctions reflect themselves in the $n(p)$ discussed in connection with Eq. 4.17. In Eq. 4.42, we give $n(p)$ for massless Bose-Einstein and Fermi-Dirac particles, respectively, when these particles are assembled in a noninteracting (ideal) gas at temperature T. Thus (the reason for the subscript i will be explained shortly)

$$n(p)_i^{\mathrm{BE}} = \frac{1}{e^{(pc+\mu)/kT} - 1} \tag{4.42a}$$

while

$$n(p)_i^{\mathrm{FD}} = \frac{1}{e^{(pc+\mu)/kT} + 1} \tag{4.42b}$$

Note the difference in the denominators. All things being equal, there are, because of the exclusion principle, fewer Fermi-Dirac particles in a given volume. Some additional remarks are in order. If the particles were massive, we would simply replace pc by $[(pc)^2 + m^2c^4]^{1/2}$ in these formulas. The quantity μ is known as the chemical potential. It arises, as we will now see, when there is a conservation law for the number of particles. In all cases the particle density n_i is given by

$$n_i = \int \frac{d^3p}{(2\pi\hbar)^3} n(p)_i \tag{4.43}$$

In the expanding universe these densities decrease as the volume expands, but it can happen, because of a conservation law, that the *total* number of particles, N_i, where

$$N_i = R^3 n_i \tag{4.44}$$

is conserved. More generally, some combination of $N_i(A)$, where A refers to some species of particle, might be conserved. In this case, the chemical potentials of these particles, the various μ_A, must readjust themselves as a function of temperature to ensure overall particle conservation. Equation 4.43 becomes the implicit equation that determines μ as a function of T. Finally, a comment about the subscript i in these formulas is in order. It allows for the fact that particles can have internal degrees of freedom such as spins or polarizations which must be summed over to find the total number of degrees of freedom these particles exhibit when they are in equilibrium. For example, the photon has two possible states of polarization that must be summed over, and spin-$\frac{1}{2}$ particles have two spin states, "up" and "down", that must be summed over.

It is a general principle of statistical mechanics that when particles are in equilibrium, their chemical potentials themselves obey a conservation law. Suppose, for example, we have particles A, B, C, and D that are undergoing the reaction

$A + B \rightarrow C + D$ and that this reaction obeys the criterion for equilibrium given by Eq. 4.3, then one can show that the respective chemical potentials μ_A, μ_B, μ_C and μ_D obey the relation

$$\mu_A + \mu_B = \mu_C + \mu_D \tag{4.45}$$

If there are more particles with chemical potentials in the initial or final state, then these too must be included in Eq. 4.45. We can use this result to show that the chemical potential of the photon in equilibrium is zero. The underlying reason is that the number of photons is not conserved in general. For example, one may have the inelastic process $e + e \rightarrow e + e + \gamma$—electron bremsstrahlung—in which photons are produced. If this process takes place under equilibrium conditions, we have

$$\mu_e + \mu_e = \mu_e + \mu_e + \mu_\gamma \tag{4.46}$$

from which it follows that $\mu_\gamma = 0$. We can use this result to argue that particles and antiparticles have equal and opposite chemical potentials. Let us call the particle B and its antiparticle $\bar{B}$. In general, one of the reactions these objects can undergo is annihilation into gamma rays: $B + \bar{B} \leftrightarrow \gamma + \gamma$. Since $\mu_\gamma = 0$, the result follows. Neutrinos can annihilate by the weak neutral current reaction $\nu + \bar{\nu} \leftrightarrow e^+ + e^-$. Since $\mu_{e^+} = -\mu_{e^-}$, from the photon annihilation argument, it follows that $\mu_{\bar{\nu}} = -\mu_\nu$.

Since $\mu_\gamma = 0$, the distribution function for each polarization of the photon, $n_{\gamma i}$, takes on the familiar blackbody form

$$n_{\gamma i} = \frac{1}{e^{pc/kT} - 1} \tag{4.47}$$

Suppose that through rapid interactions the photons have achieved this equilibrium distribution with, for example, the electrons. Furthermore, suppose that after doing so they simply go with the cosmic flow—the expanding universe—without further interactions. What happens to the distribution given by Eq. 4.47? Assuming that entropy is conserved, we must have $T \sim 1/R$. But following the discussion of Eq. 2.9, $p = h/\lambda$, the photon's momentum also decreases as $1/R$. If we call the red-shifted momentum and temperature p' and T', respectively, then we have

$$n(p, T)_{\gamma i} = \frac{1}{e^{pc/kT} - 1} = \frac{1}{e^{(pc/R)/(kT/R)} - 1} = n(p', T')_{\gamma i} \tag{4.48}$$

In other words, the relative number of photons at the red-shifted momentum and temperature is the same as the relative number of photons at the original momentum and temperature. The blackbody curve does not change its shape. However, what cosmologists do is fix a value of the momentum, say $\bar{p}$, and observe how the relative number of photons at this fixed value of p varies as a function of the temperature. This changes in time as the photons cool off. If one plots this curve for the different momenta $\bar{p}$, it has a characteristic shape at a given temperature. This is what enables us to determine the temperature of the photons.

The discussion we have just presented assumes that a blackbody spectrum has been achieved and that after this happens the photons are released to expand freely with the universe. We now want to explore the mechanisms by which this occurs. The sequence of events is a bit complicated, so we first give a general qualitative outline and then gradually fill in the details. In the qualitative outline we shall make a number of statements without justifying them, leaving the justification for the details. The key energy to keep in mind is the binding energy of the ground state of hydrogen which is, to sufficient accuracy, 13.6 eV. Why is this important? Neutral hydrogen scatters low-energy photons with great difficulty. Photons that have energies above 13.6 eV can photoionize the atom, while photons with less than this energy cannot. Hence if the cosmic photons at temperatures below a certain critical temperature related to the binding energy find themselves in an environment of neutral hydrogen, they are effectively free particles since they cannot, on the average, ionize the hydrogen. These photons will simply expand with the universe. At first sight, one might think that the photon temperature at which this photon liberation takes place would correspond to the 13.6-eV binding energy itself—about 10^5 K. However, there is a subtlety. According to the blackbody distribution formula (Eq. 4.47), there are photons of every energy at every temperature. It is a question of their relative numbers. We shall argue that until the photons cool down to about 10^4 K, there are still enough of them around with energies greater than 13.6 eV to take part significantly in the process $\gamma + \mathrm{H} \rightarrow e^- + p$—the photodisintegration of hydrogen—which keeps the neutral hydrogen from forming. Below this temperature the inverse process $e^- + p \rightarrow \gamma + H$ can take place unimpeded—and very rapidly—so that neutral hydrogen is formed and the cosmic photons are released.

But do the cosmic photons have a blackbody distribution at this temperature? The answer is yes. It turns out, as we shall see, that the elastic scattering process $e^- + \gamma \rightarrow e^- + \gamma$ takes place rapidly enough to keep the photons and electrons well equilibrated even at these low temperatures. We say "low" because there is another effect that begins to intervene at equivalent temperatures on the order of the electron's rest energy—0.51 MeV—about 10^{10} K as compared to the 10^4 K we have been talking about—that eliminates most of the electrons. This is the reaction $e^+ + e^- \rightarrow \gamma + \gamma$: the annihilation of electrons and positrons which proceeds in one direction when the photons are no longer energetic enough to make an electron-positron pair. But some electrons must remain after these annihilations have taken place since we know from experiment that the universe is sensibly electrically neutral. There must be enough electrons left over after this annihilation to balance the proton charge. There are, as far as we know, no significant amounts of antiprotons in the universe, so they can't balance the charge. The remaining electrons must do it. It turns out, as we shall see, that there are enough of them left to do the job of equilibration of electrons and photons. That is the general story. Now to the details.

We begin by discussing what cosmologists have given the unfortunate name "recombination". This is the process by which electrons and protons combine to produce hydrogen; that is, $e^- + p \rightarrow \mathrm{H} + \gamma$, where H is the neutral hydrogen atom. There is no reason to call this recombination since the electrons and protons, as far as we know, were never combined at any time prior. But that is the name you will find in all the textbooks. We are going to show that recombination takes place very rapidly at a temperature well below the one corresponding to 13.6 eV. In doing so, we will employ a great deal of the material we have learned. It will be a marvellous exercise in equilibrium statistical mechanics. We begin by noting that for all the massive particles involved, $kT \ll mc^2$. This means we can take the nonrelativistic limit of the distribution functions $n(p)_i^{\text{FD}}$ and $n(p)_i^{\text{BE}}$. We leave it to the reader to show that these two limits are the same, and we call the result $n(p)_i^{\text{cl}}$ where "cl" means classical. The nineteenth-century inventors of statistical mechanics, such as James Clerk Maxwell, were familiar with this distribution. Thus

$$n(p)_i^{\text{cl}} = e^{-p^2/2mkT} \times e^{(\mu - mc^2)/kT} \tag{4.49}$$

We can integrate this distribution to find the number density n_i^{cl}. Thus

$$n_i^{\text{cl}} = e^{(\mu - mc^2)/kT} \int \frac{d^3p}{(2\pi\hbar)^3} e^{-p^2/2mkT} = e^{(\mu - mc^2)/kT} \left(\frac{mkT}{2\pi\hbar^2} \right)^{3/2} \tag{4.50}$$

In the epoch we are discussing, all the protons were either in hydrogen, in light nuclei like helium, or free. Let us, to the accuracy that we are working, neglect the relatively small fraction of protons and electrons in the light elements. Thus we can write for the total number density of protons, n_p,

$$n_p \simeq n_{\text{free}} + n_{\text{H}} \tag{4.51}$$

where n_{free} is the number of free protons and n_{H} is the number of neutral hydrogen atoms. We can define the so-called fractional ionization X_p as

$$X_p \doteq \frac{n_{\text{free}}}{n_p} \tag{4.52}$$

The term "ionization" enters here because to every free proton there responds an electron to maintain electrical neutrality. When this fraction is 1, we have only free electrons—total ionization—and when it is 0, all the electrons have been captured into hydrogen atoms and the photons are free. If we tried to use Eq. 4.49 directly to find X_p when the reactions are in equilibrium, we would run into the problem of determining the chemical potentials. But we can get around this by remembering that in equilibrium the chemical potentials in the reaction $\text{H} + \gamma \leftrightarrow e + p$ satisfy $\mu_{\text{H}} = \mu_{\text{free}} + \mu_e$. Thus if we study the ratio

$$\frac{n_{\text{H}}}{n_{\text{free}} n_e},$$

the chemical potentials will cancel out. Furthermore, if B is the electron-proton binding energy, we have

$$m_e c^2 + m_p c^2 - m_{\text{H}} c^2 = B \tag{4.53}$$

Thus the masses will also cancel out in the exponentials, leaving only the binding energy. But there is a further subtlety. This has to do with the degrees of freedom. When we take the above ratio, we must multiply each distribution function by the number of degrees of freedom associated with each particle. We call this number g_{A}, where A is the particle in question. Hence the ratio will involve the fraction $g_H/(g_{free}g_e)$. What are these g's? The proton and electron each have spin $\frac{1}{2}$, hence for each of them g is 2. But the hydrogen atom in the ground state has four possible spin states: three corresponding to the angular momentum 1 states when the spins of the proton and electron are "parallel" and one corresponding to the angular momentum 0 state when they are antiparallel. In the hydrogen atom, the ground state has zero orbital angular momentum, so the orbital angular momentum does not contribute to the number of degrees of freedom. Thus, in this case, the value of the ratio of the g's is 1. If, in the appropriate equilibrium distribution functions, we neglect in the $mkT/\hbar^2$ terms the difference between the free proton and the hydrogen rest energies—a small correction—we will have the elegant equation

$$\frac{n_H}{n_{free} n_e} = \left(\frac{m_e kT}{2\pi\hbar^2}\right)^{-3/2} e^{B/kT} \tag{4.54}$$

But, as we have remarked, $n_{free} = n_e$. Hence we have the following equation—an example of what is known as a Saha equation—for X_p:

$$\frac{1 - X_p}{X_p^2} = \frac{n_p n_H}{n_{free}^2} = n_p \left(\frac{m_e kT}{2\pi\hbar^2}\right)^{-3/2} e^{B/kT} \tag{4.55}$$

The qualitative behavior of X_p as a function of T is determined by the rapidly varying exponential. When $kT \gg B$, the exponential tends to 1 and the expression tends to 0, corresponding to $X_p \sim 1$ or $n_p \sim n_{free}$—total ionization. On the other hand, when $kT \ll B$, the exponential is huge, corresponding to $X_p \sim 0$, which means that there are no free electrons—all have been captured to make neutral hydrogen. To make things quantitative we need an expression for n_p. We shall proceed by multiplying and dividing by n_γ where, recalling from Eq. 1.46 of the Micropaedia,

$$n_\gamma = \frac{2.4}{\pi^2}\left(\frac{kT}{\hbar c}\right)^3 \tag{4.56}$$

At the temperatures in question, after the electron-positron annihilation has taken place, the total numbers of both protons and photons are conserved. Both of these numbers, then, diminish as R^3 as the universe expands. Hence the ratio

$$\frac{n_p}{n_\gamma} \doteq \eta \tag{4.57}$$

is independent of the temperature, and we can evaluate it for the present temperature. There is some uncertainty in the value of η. We know that the number density of the photons is very close to 417 cm^{-3}. The uncertainty comes in the value of n_p. In the Micropaedia we noted that, from observation,

$$\rho_m \simeq 3 \times 10^{-7} m_p \text{ cm}^{-3} \tag{4.58}$$

which means that

$$n_p \simeq 3 \times 10^{-7} \text{ cm}^{-3} \tag{4.59}$$

Thus

$$\eta \simeq 10^{-9} \tag{4.60}$$

Much better limits on η can, as we shall see later, be derived by a study of the early universe synthesis of light elements like helium. These give $3 \times 10^{-10} \leqslant \eta \leqslant 5 \times 10^{-10}$. The crude value of η given by Eq. 4.60 will do for present purposes.

The criterion for when recombination becomes effective is somewhat arbitrary, but certainly a reasonable criterion would be to set $X_p = \frac{1}{10}$ and to solve for the critical temperature T_c. We do this using Eq. 5.55, which we rewrite by putting in the explicit form for n_γ:

$$\frac{1 - X_P}{X_p^2} = \eta \times e^{B/kT}\left(\frac{kT}{m_e c^2} \times \frac{2}{\pi^{1/3}}\right)^{3/2} \times 2.4 \qquad (4.61)$$

The curious numerical factors arise from the various definitions. To a first approximation, we can solve this equation for T_c by taking the logarithm of both sides and ignoring the $T^{3/2}$ term. We leave it as an exercise to include this term and solve the equation graphically. If we take $X_p = \frac{1}{10}$, we find $T_c \simeq 6 \times 10^3$ K, an order-of-magnitude lower temperature than the 10^5 K that corresponds to the 13.6-eV binding energy. We also leave it as an exercise for the reader to plot X_p as a function of T taking $\eta = 10^{-9}$, say. You will be surprised by how dramatically the ionization sets in and how rapidly it is completed.

As we have mentioned, when the temperature falls below about 10^{10} K, the electrons and positrons begin to annihilate. By the time recombination takes place, the positrons have all disappeared, leaving enough in the way of residual electrons to balance the proton charge. In other words, in this regime, the electron density n_e is given by

$$n_e = n_p = \eta \times n_\gamma \simeq 10^{-9}\left(\frac{kT}{\hbar c}\right)^3 \qquad (4.62)$$

The question then arises whether there are sufficient electrons left so that the process $\gamma + e \rightarrow \gamma + e$ takes place rapidly enough to keep the electrons and photons in thermal equilibrium, hence ensure the existence of a blackbody spectrum. The answer will turn out to be so clearly yes that we can get away with a very crude order-of-magnitude calculation. In the regime in question, radiation dominance is changing into matter dominance. But in the interest of simplicity, we do the calculation assuming radiation dominance. Thus we can use Eq. 4.14 in the approximate form

$$\frac{\dot{R}(t)}{R(t)} \simeq \left(\frac{kT}{m_p c^2}\right)^2 10^5\ \mathrm{s}^{-1} \qquad (4.63)$$

Let us take for kT, 1 eV, and for $m_p c^2$, 10^9 eV, which gives $\dot{R}/R \simeq 10^{-13}\ \mathrm{s}^{-1}$. The reciprocal of this gives roughly the time at which recombination takes place. We now want to compare $\dot{R}/R$ with ω, the rate for $\gamma + e \rightarrow \gamma + e$. To find this rate we need both the cross section and the flux. For the former we take simply the square of the classical electron radius—see Eq. 1.30 of the Macropaedia. This is correct to within numerical factors of order 1. Thus

$$\sigma \simeq 10^{-25}\ \mathrm{cm}^2 \qquad (4.64)$$

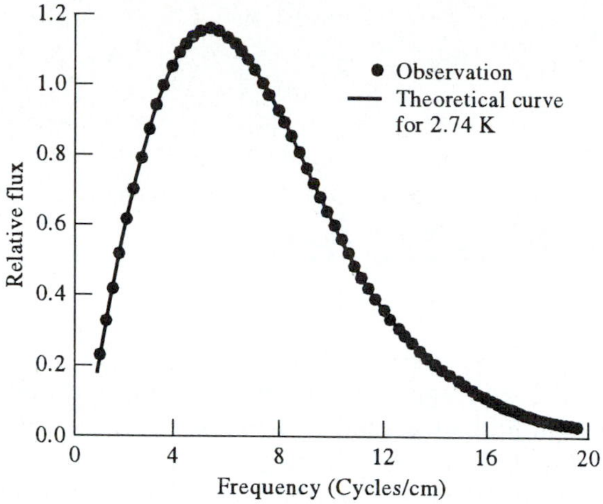

Figure 3 The observed blackbody curve.

The $\langle|\mathbf{v}|\rangle$ that occurs in the rate for electron-positron scattering is the average *relative* electron-photon speed, which is c, the speed of light. Putting everything together we have

$$n\sigma \langle|v|\rangle \simeq 7.95 \times 10^{-35}\left(\frac{kT}{\hbar c}\right)^3 c \tag{4.65}$$

If we use 1 eV for kT in Eq. 4.65, we find that $n\sigma \langle|\mathbf{v}|\rangle \simeq 3 \times 10^{-10}\,\mathrm{s}^{-1}$. Thus at 1 eV the reaction rate for elastic electron-photon scattering is three orders of magnitude greater than the rate of the universe's expansion, and the electrons and photons remain well equilibrated during the recombination era. This discussion gives us a general picture of why we expect that cosmic background radiation should have a blackbody form. What does experiment show? In Fig. 3 we present the latest data from the cosmic background explorer satellite. One could not draw a more perfect blackbody curve.

We will have a great deal of use for the ideas generated in this chapter. But in the next chapter we will switch gears entirely and discuss elementary particles.

5

Elementary Particles

If this book had been written, say, 20 yr ago, it is very unlikely that a chapter like the present one would have seemed necessary. One of the most significant advances in recent cosmology has been the discovery of the deep connections between the macroscopic and microscopic worlds. While it is true that as early as the late 1940s cosmologists realized the importance of an elementary particle like the neutrino to the workings of the early universe—the neutrino had not even been observed in the laboratory—no one, until the last few decades, realized that all the details of neutrino physics could be tested cosmologically. How many types are there? Do they have distinct antiparticles? Do they have masses? Are they stable? For this reason, to be complete in its essentials, any book on cosmology must now have a chapter on elementary particles. Since a huge book could be written about this subject, we must be selective in our choice of topics.

The broadest categorization of elementary particles is by the interactions they can have. In the early days of this subject the division was made into leptons ("light" particles such as the electron and the neutrino), mesons (particles of intermediate mass such as the pi meson), and baryons ("heavy" particles such as the protons and neutrons). The last two categories of particles had strong, weak, and electromagnetic interactions, while the leptons had only electromagnetic and weak interactions. This division is still used, although the categories often overlap. For example, for historical reasons, the so-called mu lepton, which has a mass energy of 105.658389 MeV, is always referred to as the mu meson, while the so-called tau lepton with its mass energy of 1784.1 MeV—its mass is known with much less accuracy—has a mass that is nearly two times as great as that of the proton, which is known as a baryon. A glance at a table of elementary particles is enough to render the uninitiated a little dizzy. In a recent one, for example, there are about 115 pages devoted to a listing of these particles and their properties. Fortunately, there are regularities among the particles which reflect symmetry properties of their interactions, and these symmetries render some order in the apparent chaos. The first such symmetry property to have been recognized among the strongly interacting particles was what was called isotopic spin symmetry. It was introduced in the 1930s. Figure 1a and b suggest how the existence of this symmetry was discovered.

What is striking is that the mass differences between, for example, the proton and the neutron, or the three pi mesons, are very small compared to the masses

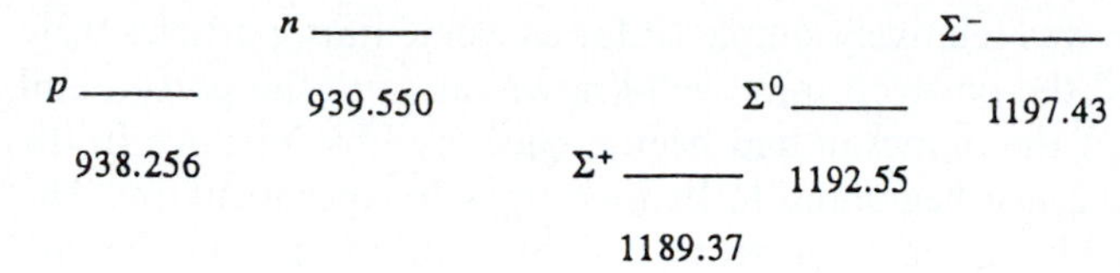

Figure 1a Some baryons with their mass energies in megaelectronvolts.

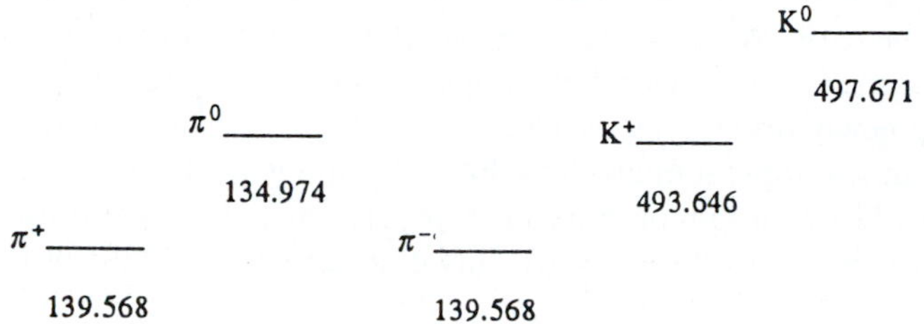

Figure 1b Some mesons with their mass energies in megaelectronvolts.

themselves. In fact, recalling that $\alpha = e^2/4\pi\hbar c \simeq 1/137$, we see that these mass differences are all of the order of αm, where m is the mass in question. This suggests that in a hypothetical world in which electromagnetism could be switched off—in which $\alpha = 0$—these mass differences might disappear. The particles might then form mass degenerate multiplets as shown in Fig. 2a and b. We have also shown the isotopic spins assigned to these multiplets and the z components of these spins.

Notice the relation between the electric charges of the particles in the left column of the figures and the z components of their isotopic spins, namely,

$$Q = \frac{B}{2} + T_z \tag{5.1}$$

Here B stands for the so-called baryon number which is 1 for p and n and 0 for the three pi mesons. This isotopic spin is conserved in strong interactions and violated in electromagnetic interactions.

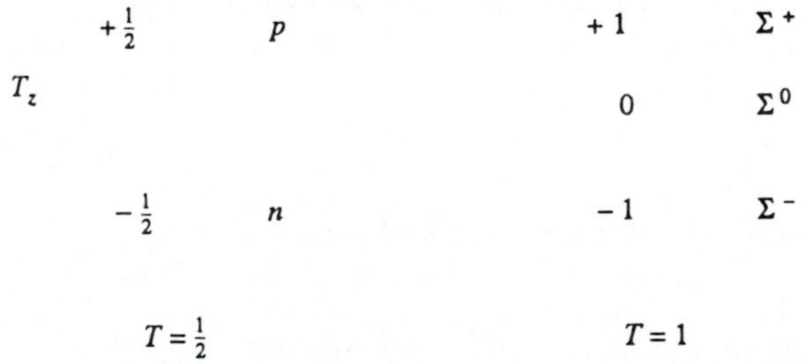

Figure 2a Some baryon isotopic spin multiplets.

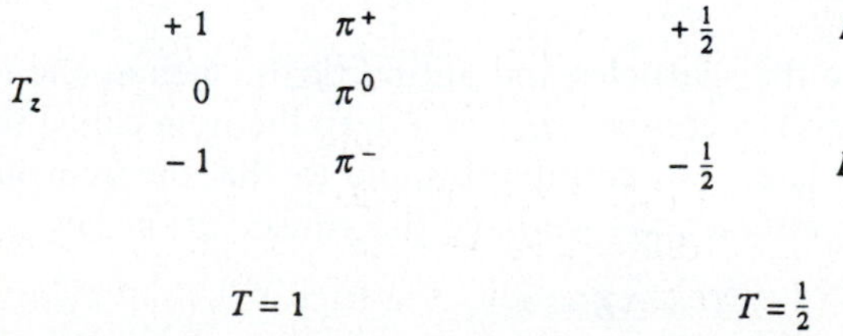

Figure 2b Some meson isotopic spin multiplets.

Until the late 1940s life was relatively simple as far as elementary particles were concerned. The electron and the positron were well-known, as were the proton and the neutron. The existence of the pi meson had been conjectured by Yukawa in the 1930s, and its properties were just beginning to be investigated experimentally. The mu meson, which behaved like a heavy electron, had been discovered in the late 1930s, although no one knew why it existed. The neutrino had been invented, for reasons we will discuss, in the early 1930s by Pauli, but had not yet been observed. And then the roof fell in. Elementary particles began appearing everywhere, first in cosmic rays and then in the newly developing accelerators. It became clear that the connection between charge and isotopic spin given by Eq. 5.1 did not work for them, at least with the conventional definition of baryon number, so a new quantum number called strangeness, S, was introduced. With this quantum number Eq. 5.1 was rewritten

$$Q = \frac{B+S}{2} + T_z \tag{5.2}$$

Now if the K mesons are assigned a strangeness of $+1$ and the Σ particles a strangeness of -1, the relationship again works. However, one can also redefine the baryon number. One introduces what is known as the hypercharge, Y, with the definition

$$Y = B + S \tag{5.3}$$

Thus

$$Q = \frac{Y}{2} + T_z \tag{5.4}$$

In Fig. 3, we present a table with hypercharge assignments. We also introduce some additional particles along with their mass energies.

Y	T	Baryons	Mesons
$+1$	$\frac{1}{2}$	p, n	K^+, K^0
0	1	$\Sigma^+, \Sigma^0, \Sigma^-$	π^+, π^0, π^-
0	0	$\Lambda^0(1115.63)$	$\eta^0(548.8)$
-1	$\frac{1}{2}$	$\Xi^0(1314.9)\Xi^-(1321.32)$	$\bar{K}^0(497.671)K^-(493.646)$

Figure 3 A table with the hypercharge and isotopic spin assignments for some of the light baryons and mesons.

The alert reader will notice that particles and antiparticles, such as the π^+ and the π^-, have the same mass. This is a consequence of a deep theorem called the CPT theorem which we will discuss later. No counterexample to this theorem has ever been discovered. In the strong interactions we have the conservation law

$$\Delta Y = \Delta B = 0 \tag{5.5}$$

But in the weak interactions there can be decays in which $\Delta Y \neq 0$. Some of the newer theories of elementary particles also predict that there can be superweak interactions in which $\Delta B \neq 0$. This would mean that the proton would be slightly unstable. Experiments have been performed to study this hypothetical proton decay. A typical decay mode that has been looked for is $p \rightarrow e^{+} + \pi^{0}$. No such decay has ever been observed. According to these experiments the partial proton lifetime for this mode is greater than about 10^{32} yr! Keep in mind that the age of the universe is "only" about 1.5×10^{10} yr.

When these particles began to proliferate, it was natural to try to introduce a simplifying scheme by which to classify them. The first thing that came to mind was the simplifying scheme for classifying atomic nuclei. If we count all the isotopes, there are hundreds of such nuclei. But we can grasp their essential nature once we realize that they are all made up of neutrons and protons. We can demonstrate this by breaking up a nucleus and studying its elementary components. It naturally occurred to people to try to do the same sort of thing with elementary particles. The first such scheme was invented by Fermi and C. N. Yang. It was a modest attempt to understand the pi mesons. They realized that the quantum numbers of the three pi mesons were the same as those of bound nucleon-antinucleon singlet, zero orbital angular momentum states. Symbolically we can write, for example, $\pi^{+} \sim p\bar{n}$. With the discovery of the strange particles it became clear that this scheme would have to be extended. One might try adding the Λ^{0} and its antiparticle. This did not quite work. There then followed a period of some desperation which lasted until the early 1960s when Murray Gell-Mann and George Zweig invented what became known as the quark model. (Zweig called the fundamental constituents "aces", but that name did not stick.) In creating this model Gell-Mann and Zweig gave up the idea that the fundamental constituents need necessarily be objects we observe directly. If present ideas are right, no one will ever be able to observe a free quark. What counts is the model's utility.

Let us temporarily use the notation q for a quark and let us see what properties these objects must have. We will begin with the three Σ particles. To construct the charged sigmas we evidently need charged quarks. Let us postulate the existence of a positively and a negatively charged quark: q^{+} and q^{-}. We might then try to construct the Σ^{+} and the Σ^{-} from three quarks as follows:

$$\Sigma^{+} \sim q^{+}q^{+}q^{-}$$

$$\Sigma^{-} \sim q^{-}q^{-}q^{+}$$

But if we are also going to make up the Σ^{0} out of three quarks, we will need a neutral quark q^{0}. We might then have

$$\Sigma^{0} \sim q^{+}q^{-}q^{0}$$

This is reasonably conventional. But what about baryon number? The Σ's have baryon number 1. This means that the quarks must have baryon number $\frac{1}{3}$! That is radical. Furthermore, the Σ's are strange. This means that there must be at least one strange quark. But it can't be neutral since both the charged and the neutral Σ's carry strangeness. We seem to be stuck. However, while we are at it, there is no reason to assume that the q^+ and the q^- carry equal and opposite charge. Let us try again with three quarks that we now call u (up), d (down), and s (strange). We will be open-minded about their eventual charges and write

$$\Sigma^+ \sim uus$$

$$\Sigma^0 \sim uds$$

$$\Sigma^- \sim dds$$

We may now produce three equations for the quark charges, namely,

$$\begin{aligned} 2q_u + q_s &= 1 \\ q_u + q_s + q_d &= 0 \\ 2q_d + q_s &= -1 \end{aligned} \tag{5.6}$$

However, these equations are not independent. Adding the first and last produces the middle one. We need at least one more equation to determine the charges. To this end we can consider the Ξ^0. From Fig. 3 and Eq. 5.3, it has strangeness -2. We can try

$$\Xi^0 \sim uss$$

This gives us an additional equation

$$q_u + 2q_s = 0 \tag{5.7}$$

Now the equations for the q's do have a solution, namely,

$$\begin{aligned} q_u &= \tfrac{2}{3} \\ q_s &= -\tfrac{1}{3} \\ q_d &= -\tfrac{1}{3} \end{aligned} \tag{5.8}$$

These are the famous fractional charges of the quarks. When Gell-Mann first invented them, it was an open question of whether or not they existed as free particles. A search was undertaken which explored everything from seawater to moon rocks—all to no avail. By now this lack of success has been elevated to a principle, the principle of "confinement". The quarks in elementary particles are permanently confined there and will never get out. All experiments done so far on elementary particles are consistent with this principle.

From our three quarks it is easy to see that we can construct 10 distinct kinds of baryons. When the scheme was first proposed, not all of these baryons had been found. Discovering them was a real triumph for the model. Figure 4 is a full table that includes masses for the other particles: the four Δ's and the Ω^-. These particles are interesting because they have spin $\frac{3}{2}$, unlike the other baryons that we have already been introduced to. It is clear that to produce spin $\frac{1}{2}$ or spin $\frac{3}{2}$ particles from three quarks they must themselves have half odd-integer spin. All the evidence is consistent with assigning them spin $\frac{1}{2}$. This means that they obey the Pauli exclusion principle, a point that we will have occasion to return to. We could also imagine binding the quarks in states with nonzero orbital angular momenta. We would expect such states to be less tightly bound, and indeed particle tables are festooned with such unstable objects.

Quark content	Strangeness	Charge	Particle	
uuu	0	+2		$\Delta^{++}(1232)$
uud	0	+1	*p*	$\Delta^{+}(1232)$
udd	0	0	*n*	$\Delta^{0}(1232)$
ddd	0	−1		$\Delta^{-}(1232)$
uus	−1	+1	Σ^+	
uds	−1	0	Σ^0	Λ^0
dds	−1	−1	Σ^-	
uss	−2	0	Ξ^0	
dss	−2	−1	Ξ^-	
sss	−3	−1	Ω^-	

Figure 4 The quark model of the light baryons.

Before constructing a similar table for the light mesons we have to discuss antiquarks. Antiquarks necessarily enter here because mesons have baryon number 0. The baryon number of the quarks must cancel the antibaryon number of the antiquarks. It is natural to give isotopic spin assignments to the quarks. Otherwise, the quark model would give us no framework within which to fit the isotopic spins of the observed particles. Figure 5 shows both the isotopic spins and the hypercharges for the quarks. The latter are from Eq. 5.3.

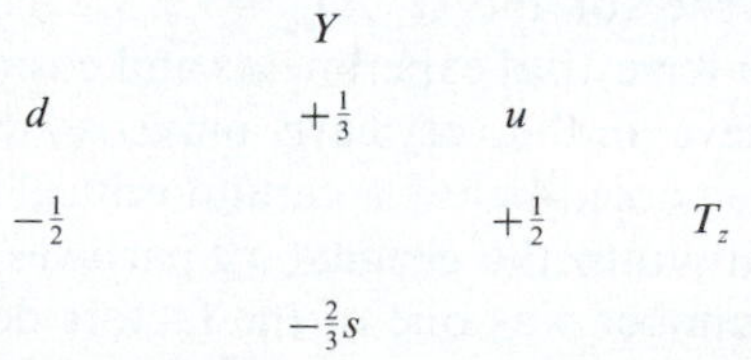

Figure 5 The quarks with their isotopic spin and hypercharge assignments.

The antiquarks have equal and opposite charges to the quarks. The baryon numbers are also flipped. This enables us to construct the corresponding antiquark table (Fig. 6).

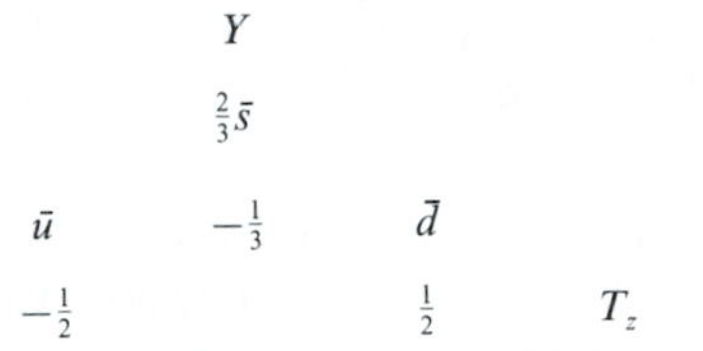

Figure 6 Antiquarks with their isotopic spin and hypercharge assignments.

Figure 7 lists the nine distinct combinations of quarks and antiquarks that produce the observed spin-0 mesons. We also give the mass energies for the spin-0 mesons that we have not introduced before. There is, in addition, a set of nine spin-1 mesons, all of which have been observed, corresponding to the quarks and antiquarks being bound with their spins parallel. We will not give the table of these spin-1 mesons since they do not enter our cosmological story.

Quark content	Strangeness	Charge	Particle
$u\bar{s}$	+1	+1	K^+
$d\bar{s}$	+1	0	K^0
$u\bar{d}$	0	+1	π^+
$d\bar{d}$, $u\bar{u}$, $s\bar{s}$	0	0	π^0, η^0, $\eta^{0\prime}(958)$
$d\bar{u}$	0	−1	π^-
$s\bar{u}$	−1	−1	K^-
$s\bar{d}$	−1	0	$\bar{K}^0$

Figure 7 The quark model of the light spin-0 mesons. Note that the neutral strangeness spin-0 mesons are made up of various combinations of the indicated quarks and antiquarks. We have not exhibited these combinations explicitly since they will not arise in our later work.

It was understood rather early in the quark game that there was a potential problem with the model. It shows itself clearly in the case of the Ω^-. As we have seen, from the point of view of the quantum numbers, we can think of the Ω^- as being made up of three s quarks in a state of zero orbital angular momentum. Since the Ω^- has spin $\frac{3}{2}$, one would be tempted to put all three s quarks in the same spin state. But this conflicts with the Pauli exclusion principle. Once this was realized, various ways out were proposed. Now it is generally agreed that the correct way out is to suppose that the quarks are triplicated. In other words, that each flavor of quark—u, d, s—comes in three "colors"—say red, white, and blue. This does not mean that the quarks are actually painted. It is only a mnemonic to remind us of the triplication. Thus, for example, we would write for the Ω^-, $\Omega^- \sim s_r s_w s_b$. This triplication has observable consequences both for terrestrial experiments and cosmology. As we shall discuss, there was an era, we believe, in the very early universe, where there actually *were* free quarks. As the universe cooled below a certain critical temperature, these quarks became bound permanently into the elementary particles we observe today. But while they were free, their number was one of the factors determining the rate at which the universe expanded. We can understand this qualitatively since the energy density of the universe depends on the number of particles at our disposition and the expansion rate depends on the energy density. But the energy density evidently depends on the number of particles, hence the expansion rate of the universe during this epoch depends on the number of quarks. We will come back to these matters more fully in a later chapter.

The picture with three flavors of quarks began coming unstuck in the mid-1970s. New mesons began to appear which did not fit into the three-flavor tableaux. For example, there are various so-called strange D mesons with masses in the 2-MeV range. To fit these, a new flavor of quark was introduced and given the unfortunate name of "charmed" quark c. One can then construct combinations like $c\bar{s}$. One can also construct nonstrange charmed mesons with combinations like $c\bar{u}$. All of these have also been observed. But this was not the end of it. Soon after, another set of mesons was found that needed yet another quark which was given the even more unfortunate name of "bottom" quark, b. There is good reason to think it has a partner, the "top" or t quark, its mesons may recently have been observed. On its face, one might think that this proliferation of flavors could go on forever. However, remarkably, we may have seen the end of it. The evidence for this is both terrestrial and cosmological. Indeed, the whole point of this chapter is to prepare ourselves to understand the cosmological evidence. Before getting to this we will need to explore the world of the leptons and, above all, the world of the neutrino.

The neutrino was invented by Pauli in 1930. He invented it in response to a puzzle involving β decay—the decay of an elementary particle or an atomic nucleus in which an electron or a positron is emitted. For historical reasons, in weak interactions, the electron or positron is often referred to as a β particle. The puzzle had to do with the spectrum of energies with which these particles emerged from the decay. In relativity theory we write the energy of a free particle as

$$E = \sqrt{p^2c^2 + m^2c^4} = T + mc^2 \tag{5.9}$$

an equation that defines the relativistic kinetic energy T. Experimenters then plot the relative numbers of electrons or positrons that emerge from β decays as a function of T, the quantity $N(T)$. Figure 8 presents a plot of $N(T)$ as observed in a typical β decay.

The most striking thing about this plot is that there is a continuous spectrum. The reason that this is striking has to do with the kinematics of the decay. Suppose the decay were of the form $A \rightarrow B + \beta$—a two-body decay. Then, and we leave this as an exercise for the reader, the conservation of energy and momentum would determine T uniquely as a function of the masses. Indeed, in a two-body decay of the form $A \rightarrow B + \beta$

$$T = \frac{(m_A^2 + m_\beta^2 - m_B^2)c^2}{2m_A} - m_\beta c^2 \tag{5.10}$$

This means that instead of being a continuum of energies, the curve in Fig. 8 would degenerate into a single point.

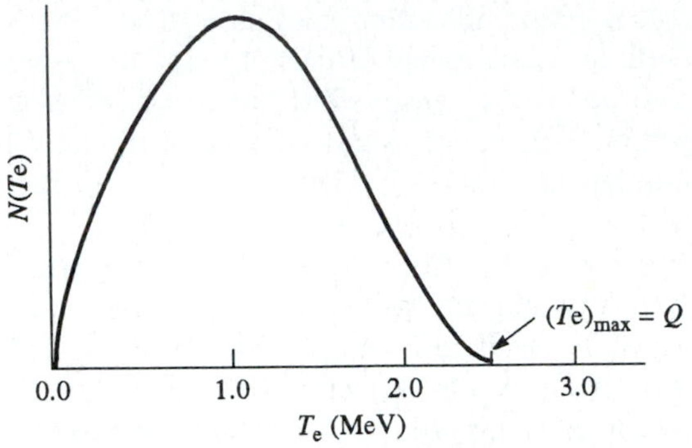

Figure 8 The β-energy spectrum produced in a typical decay.

When this was realized in the late 1920s, a certain amount of panic set in since in beta decays only two outgoing particles were observable: the beta particle and the daughter nucleus. Indeed, Bohr even suggested that energy conservation might be violated in these decays. To save the situation, Pauli suggested that there was an invisible third particle emitted along with the other two—an X particle. Then the decay would be of the form $A \rightarrow B + \beta + X$. If one re-does the kinematics of this three-body decay, one will discover that only the maximum and minimum values of T are fixed. The minimum value is zero, and the maximum value, $T_{\max}$, is given by

$$T_{\max} = \frac{[m_A^2 + m_\beta^2 - (m_B + m_X)^2]c^2}{2m_A} - m_\beta c^2 \qquad (5.11)$$

The form of the spectrum is not determined from the kinematics but rather from the quantum dynamics of the decay.

A prototypical β decay is the decay of the neutron: $n \rightarrow p + e^- + X$. Since the neutron is electrically neutral, Pauli reasoned that the X had to be electrically neutral. From the kinematics Pauli argued that its mass had to be less than the mass of the electron. The spin of the X was more subtle since, when Pauli made his proposal, the neutron itself had not yet been observed—only conjectured—and there was some confusion about the spin and statistics of nuclei. Nonetheless, Pauli's physical intuition saw him through, and he made the correct hypothesis that the X had spin $\frac{1}{2}$. The task of using the X in a theory of β decay was left to Fermi. He made the first calculation of the spectrum using four-Fermi interaction, whose picture we have shown more than once. By this time the neutron had been observed. It was called the *neutrone*—the "big neutral one" in Italian—so Fermi called the X the *neutrino*—the "little neutral one"—and that name has endured. For the next 30 yr, people developed the neutrino theory of β decay and made innumerable measurements without anyone having observed an actual neutrino. This changed radically in 1956 when G. Cowan and F. Reines finally observed neutrinos, or rather, antineutrinos. We will return later to how neutrinos and antineutrinos can be distinguished, but first let us explain the Cowan–Reines experiment. What made this experiment possible was the existence of giant nuclear reactors. The one that Cowan and Reines used was located at the Savannah River plant in South Carolina which houses a uranium fission reactor. In these reactors unstable fission products are produced—radioactive nuclei in the middle of the periodic table. These nuclei are "neutron-rich"—they contain more neutrons than protons. They turn themselves into stable nuclei in which the number of neutrons and protons equalize through electron β decays. By convention, the electron is called the particle, and the positron the antiparticle. Hence, to balance things, the object that is emitted with an electron in β decay is by convention called the antineutrino $\bar{\nu}$, while the object that is emitted with the positron is designated as the neutrino ν. We can formalize this arrangement by introducing a so-called lepton number 1. This number is then assumed to be conserved in these decays. Figure 9 lists the lepton numbers of the particles we have been discussing.

Particle	Lepton no.	Particle	Lepton no.
p	0	e^-	+1
$\bar{p}$	0	e^+	−1
n	0	ν	+1
$\bar{n}$	0	$\bar{\nu}$	−1

Figure 9 The lepton numbers of some of the particles.

Notice that both the reactions $n \rightarrow p + e^- + \bar{\nu}$ and $\bar{\nu} + p \rightarrow e^+ + n$ conserve lepton number, while the reaction $\bar{\nu} + n \rightarrow e^- + p$ violates lepton number conservation. The Savannah River reactor produced a stupendous antineutrino flux of about $5 \times 10^{13}\ \text{cm}^{-2}\ \text{s}^{-1}$. Some of these antineutrinos impinged on a specially constructed tank about 3 in. high and $6\frac{1}{4}$ by $4\frac{1}{4}$ ft in area, placed at a safe distance from the reactor. It held about 2 liters of a water–cadmium acetate mixture. The reason for the latter will become evident shortly. The problem with neutrino reactions is that there are very few of them. A neutrino can penetrate 1000 *light-years of lead* before it interacts since it does so only by weak interactions. Hence one needs a "signature" for a neutrino interaction that stands out like a sore thumb. This is the ingenuity of the Cowan-Reines arrangement. In Fig. 10 we present a kind of cartoonlike series of drawings that illustrate the time sequence involved. Then we give a verbal description.

The idea is this. The tank we have described constitutes the middle of a sandwich whose upper and lower parts are composed of a scintillator—a material that scintillates (emits light)—when bombarded by radiation. This light can be detected with photomultipliers. There are both liquid and solid scintillating materials, and Cowan and Reines used liquid scintillators which were viewed by an array consisting of a large number of photomultiplier tubes. At $t = 0$ an antineutrino enters the tank. Let us suppose that it is one of the few that interacts. Indeed, let us suppose that at $t = 0^+$ it undergoes the reaction $\bar{\nu} + p \rightarrow e^+ + n$. This means that a positron is created. But it can interact electromagnetically. In particular, it can annihilate with an electron—the reaction $e^+ + e^- \rightarrow \gamma + \gamma$. This happens in about 10^{-9} s and produces two 0.51 MeV gamma rays. In the events we are looking for, these gamma rays enter the upper and lower scintillators at essentially the same time. In the

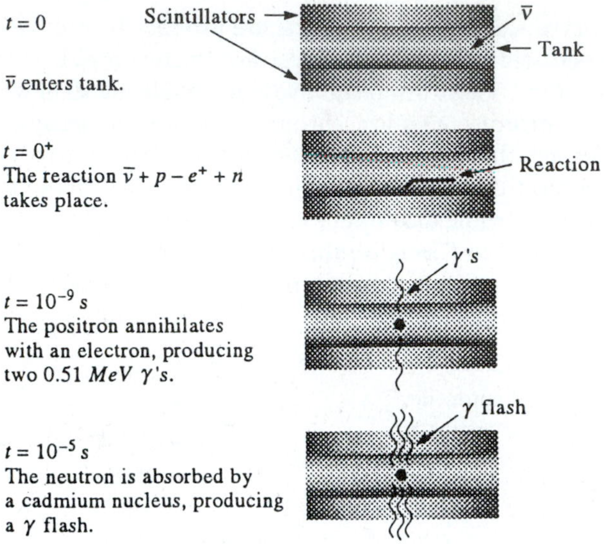
t = 0
v̄ enters tank.
Scintillators
v̄
Tank
t = 0⁺
The reaction v̄ + p – e⁺ + n takes place.
Reaction
γ's
t = 10⁻⁹ s
The positron annihilates with an electron, producing two 0.51 MeV γ's.
γ flash
t = 10⁻⁵ s
The neutron is absorbed by a cadmium nucleus, producing a γ flash.

Figure 10 The sequence of events in the Cowan–Reines experiment.

meantime, the neutron produced in the original reaction has been wandering around the tank. Cadmium is a very strong absorber of neutrons, and in about 10^{-5} s our neutron gets absorbed, accompanied by a flash of gamma rays which are detected by the scintillators. It is this timed sequence of events that stands out like a sore thumb. Despite the huge flux, Cowan and Reines counted only about three such sequences per hour. We can use this number to produce an estimate of the cross section for the original weak interaction. Remember that the rate ω is related to the flux $\mathscr{F}$ by the equation $\omega = \mathscr{F}\sigma$, where σ is the cross section. But this would be the rate produced by the scattering from a single particle. If there are N such particles, then the observed rate ω_{obs} is given by

$$\omega_{\text{obs}} = N\mathscr{F}\sigma \tag{5.12}$$

The volume of the tank is about $\frac{2}{10}$ m^3. Remembering that there are about 2.7×10^{25} nucleons per cubic meter under standard conditions, we can substitute the known $\mathscr{F}$ and ω_{obs} into Eq. 5.12 and solve for σ. The result is that σ for the reaction $\bar{\nu} + p \rightarrow e^+ + n$, as measured in this experiment, is about 10^{-42} cm^2. Is this number reasonable? In the previous chapter we gave a dimensional analysis that led to an expression for this cross section—see the discussion leading to Eq. 4.6. There we used the fact that in a thermal bath the average energy is about kT. But here we are dealing with a different situation. These antineutrinos are produced when the radioactive isotopes decay. The antineutrinos have, typically, an energy of a few megaelectron-volts. Let us say 10 to keep the numbers simple. The cross section in terms of this energy, using the same sort of dimensional analysis, is given by

$$\sigma \simeq \frac{G_F^2 E^2}{(\hbar c)^4} \tag{5.13}$$

If we use the values for G_F, $\hbar$, and c from the Micropaedia and 10 MeV for the energy, we find $\sigma \simeq 10^{-42}\,\mathrm{cm}^2$, in good agreement with the Cowan–Reines result. We can compare this number to $r_0^2 \simeq 10^{-25}\,\mathrm{cm}^2$, which is the order of magnitude we used for the cross section for $\gamma + e \rightarrow \gamma + e$. The cross sections for nuclear reactions are substantially larger, which illustrates the strength hierarchy of the three fundamental interactions. In a subsequent experiment Cowan and Reines tried to produce electrons in the reaction $\bar{\nu} + {}^{37}\mathrm{Cl} \rightarrow e^- + {}^{37}\mathrm{Ar}$—the transmutation of chlorine into argon by an antineutrino with the production of an electron—a violation of lepton number conservation. They found no events, an important confirmation of the conservation of lepton number.

No account of neutrino physics would be complete without some discussion of the "glorious revolution of 1955–1956". For those of us who witnessed it, it was a revolution. To understand it we give a little background. After the invention of the quantum theory in the 1920s it was discovered how to incorporate symmetries into the theory. These correspond to quantum mechanical operators that commute with the Hamiltonian of the system. For example, if the operator representing the orbital angular momentum commutes with a given Hamiltonian, then the orbital angular momentum is conserved. This symmetry can be related to the continuous rotations of the coordinates used to describe the system. In addition, there can be discrete transformations under which the Hamiltonian is invariant. An example is the so-called parity transformation. We illustrate what is involved in Fig. 11. It depicts the sequence of transformations needed to make a right-handed coordinate system into a left-handed one. In three dimensions, at least one of these transformations

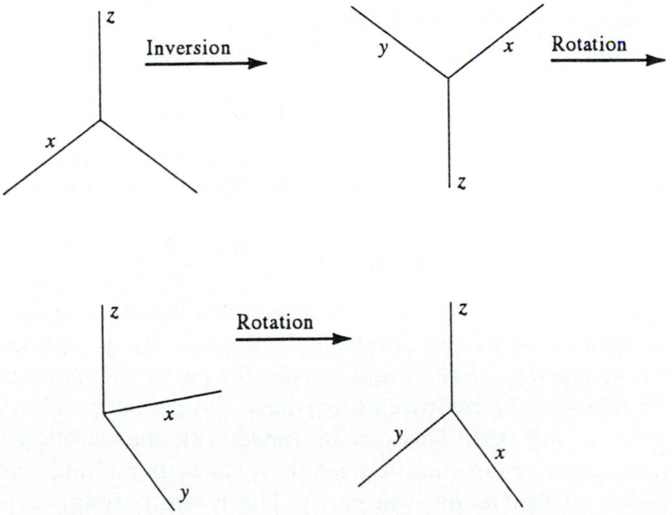

Figure 11 A sequence of transformations that changes a right-handed coordinate system into a left-handed one. Note the reflection.

must be a reflection through the origin, and this cannot be represented by a rotation through an angle. See Fig. 11.

Under the parity transformation many of the familiar quantities undergo well-defined changes. We give several illustrations from classical physics:

$$\mathbf{r} \rightarrow -\mathbf{r} \tag{5.14}$$

$$\nabla \rightarrow -\nabla \tag{5.15}$$

Thus Newton's law

$$-\nabla V(\mathbf{r}) = m\ddot{\mathbf{r}} \tag{5.16}$$

becomes

$$\nabla V(-\mathbf{r}) = -m\ddot{\mathbf{r}} \tag{5.17}$$

Hence if $V(\mathbf{r}) = V(-\mathbf{r})$, it becomes a matter of indifference whether we use right-handed or left-handed coordinates. We can extend these ideas to electromagnetism. Here we deal with the electric field $\mathbf{E}(\mathbf{r})$ and the magnetic field $\mathbf{H}(\mathbf{r})$. Under parity, a conventional vector transforms like $\mathbf{r}$. But we must also remember to transform the coordinate at which the field is being evaluated. Thus

$$\mathbf{E}(\mathbf{r}) \rightarrow -\mathbf{E}(-\mathbf{r}) \tag{5.18}$$

and

$$\nabla \times \mathbf{E}(\mathbf{r}) \rightarrow \nabla \times \mathbf{E}(-\mathbf{r}) \tag{5.19}$$

The quantity $\nabla \times \mathbf{E}(\mathbf{r})$ transforms, therefore, not like a normal vector but rather like a pseudovector. The orbital angular momentum $\mathbf{L} = \mathbf{r} \times \mathbf{p}$ also transforms like a pseudovector, as does the spin $\boldsymbol{\sigma}$. The fact that $\nabla \times \mathbf{E}(\mathbf{r})$ transforms like a pseudovector implies, if parity is to be conserved by electromagnetism, that $\mathbf{H}(\mathbf{r})$ must also transform like a pseudovector, as we can see from the Maxwell equation

$$\nabla \times \mathbf{E}(\mathbf{r}) = \frac{1}{c}\,\mathbf{H}(\mathbf{r}) \tag{5.20}$$

Quantum mechanical systems obey the Schrödinger equation

$$\left(\frac{-\hbar^2}{2m}\nabla^2 + V(\mathbf{r})\right) = i\hbar\frac{\partial}{\partial t}\psi(\mathbf{r}, t) \tag{5.21}$$

If $V(\mathbf{r}) = V(-\mathbf{r})$, we see that this equation is invariant under the parity transformation. In this case we can always find solutions that have both definite energies and definite symmetry—even or odd—under the parity transformation. For example, the energy levels of the hydrogen atom have this property. The S states are even, and the P states are odd. This can be tested with the selection rules that govern the electromagnetic transitions between these states. In addition, elementary particles can have what is called an intrinsic parity. This is mostly simply explained by the example of the spinless mesons such as the pi mesons. A pi meson is described by a field variable $\phi(\mathbf{r}, t)$. When we say that such a meson has an intrinsic parity, we mean that $\phi(-\mathbf{r}, t) = \pm\phi(\mathbf{r}, t)$. If the plus sign holds, we call the particle a scalar. If the minus sign holds, we call it a pseudoscalar. This is not a distinction without a difference. An interesting case in point is the π^0. Its most important decay mode is $\pi^0 \rightarrow \gamma + \gamma$. Using the polarization vectors of the two photons $\boldsymbol{\epsilon}_1$ and $\boldsymbol{\epsilon}_2$ and the common momentum of the photon **k**—we assume that the π^0 decays at rest so that the two photons have equal and opposite momenta—we can construct two kinds of final states that meet all the symmetry conditions, namely, something proportional to $\boldsymbol{\epsilon}_1 \cdot \boldsymbol{\epsilon}_2$ or something proportional to $\boldsymbol{\epsilon}_1 \times \boldsymbol{\epsilon}_2 \cdot \mathbf{k}$. The first quantity is a scalar, and the second a pseudoscalar. If the π^0 is, say, a pseudoscalar and if parity is conserved in the decay, which we believe it is, then the first expression is ruled out. This implies that the two photons emerge with their polarization vectors at right angles (Fig. 12). This can be tested, and experiments show that the π^0 is indeed a pseudoscalar particle.

Until 1956 nearly everyone was sure that all interactions conserved parity. But there were some odd results that began appearing in the decays of elementary particles such as the K mesons that caused some people to begin to wonder. In particular, T. D. Lee and C. N. Yang made the remarkable discovery that this symmetry had never been tested in weak interactions. People only thought that it had been. They proposed a number of experiments which were forthwith carried out. A striking one was the β decay of ^{60}Co, an isotope of cobalt. This isotope has a nonzero angular momentum which can be represented by the pseudovector **I**. These nuclei can be aligned to an external magnetic field. Their angular momenta can be made to line up with the field. When the ^{60}Co decayed, the experimenters discovered that the electrons were emitted preferentially in a direction opposite that of the aligned **I**. This implies that there is a correlation of the form $\mathbf{I} \cdot \mathbf{p}$, where **p** is the momentum of the electron. But this correlation is a pseudoscalar and cannot appear in a beta-decay theory that is entirely parity-conserving. Until Lee and Yang did their work no one had thought to measure such correlations. Discovering them created a sensation.

Among the consequences was the resurrection of a theory of the neutrino that had been invented, for its beauty, by the mathematician Hermann Weyl and then

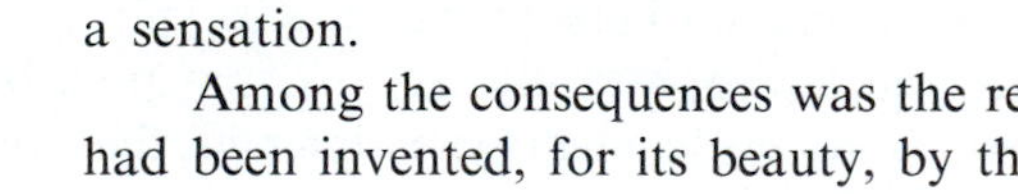

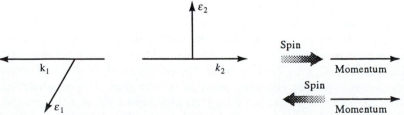

Figure 12 The configuration of the photons emitted in the decay of the π^0.

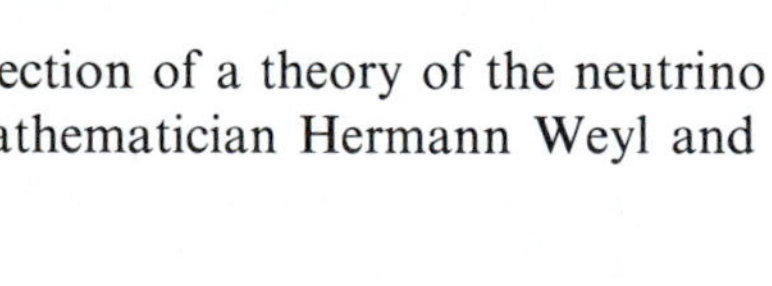

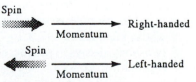

Figure 13 Neutrino spins either correlated or anticorrelated with their momenta. Note the designations "left-handed" and "right-handed".

abandoned precisely because it appeared to violate parity symmetry! In this theory the spin of the neutrino can be either correlated or anticorrelated with its momentum. In Fig. 13 we illustrate this situation.

This theory differs completely from the theory of the electron, where there is no such correlation. It depends essentially on the assumption that the neutrino is massless and therefore moves with the speed of light. Otherwise, we could outrace the neutrino and reverse its momentum, leaving its spin unchanged. We could change a left-handed neutrino into a right-handed neutrino by a change of coordinates. If the neutrino has a small mass, Weyl's theory could still be a useful approximation and, indeed, it does seem to work very well. It becomes a matter of experiment to determine whether the neutrino is right-handed or left-handed, and that is what we turn to now. This will also give us a chance to introduce the notion of neutrino families.

To do this we return to the pi mesons, but this time to the charged ones. The charged pi mesons have as their principal decay mode decay into a mu meson and some sort of neutrino or antineutrino. The lifetime of this decay for both the π^+ and the π^- is 2.60×10^{-8} s. The fact that particle and antiparticle have the same decay rate is also a consequence of the CPT theorem alluded to earlier. Note, by the way, that the π^0 lifetime is 8.4×10^{-17} s. This is in part a reflection of the fact that the π^0 decays by electromagnetic interactions, while the charged pion decay by weak interactions. Since the pi meson has spin 0, the total angular momentum of the final state must also be zero. If we assume the Weyl theory of the neutrino is right, the neutrino will have its spin lined up with its momentum or lined up opposite its momentum. In either case the mu meson must have *its* spin suitably lined up so as to conserve angular momentum. For the sake of definiteness let us consider the decay $\pi^- \rightarrow \mu^- + \bar{\nu}$. We have assumed that the mu meson is the "particle" and, in the spirit of lepton conservation, designated the object emitted with it in this decay the antineutrino. Figure 14 is a diagram that illustrates the situation assuming that the antineutrino is a right-handed particle.

Since the neutrino is very elusive, we cannot hope to measure its spin direction directly. But the mu meson interacts electromagnetically, and its spin polarization can be directly measured. When this is done, the handedness assignment that we have shown in Fig. 14 is confirmed. If one studies the decay $\pi^+ \rightarrow \mu^+ + \nu$, one can confirm that the neutrino is left-handed. This then is a sharp distinction between neutrinos and antineutrinos—their handedness.

About the time that these experiments were being performed a paradox began emerging in the theory of muon decays. The mu meson is an unstable particle. The μ^-, for example, decays via $\mu^- \rightarrow e^- + \nu + \bar{\nu}$, a lepton number-conserving decay whose lifetime is 2.197×10^{-6} s. In general, three-body decays are less probable than two-body decays. The question then arose as to why there was no observed decay of the form $\mu^- \rightarrow e^- + \gamma$. No such decay has ever been seen, and the present upper limit is at most 1 such decay per 10^{11} normal decays. When there is an absence of a process in the quantum theory which is allowed by the usual modalities of energy and momentum conservation, there is, in general, a reason, especially if one can draw a perfectly good Feynman diagram (see Fig. 15) that allows the process to take place.

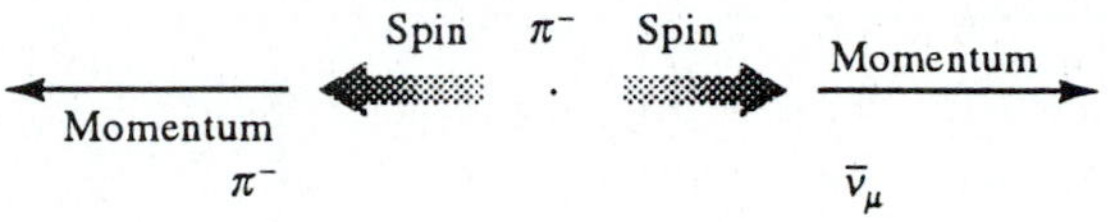

Figure 14 The decay of the π^-. Note the alignment of the spins.

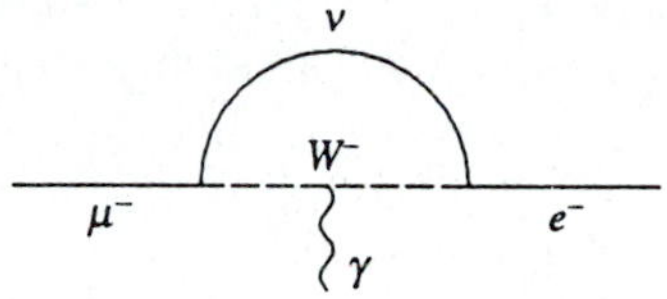

Figure 15 A Feynman diagram that leads to the process $\mu^- \rightarrow e^- + \gamma$.

Since there was no obvious selection rule to forbid the decay, a new quantum number called the "mu-ness" or muon number was introduced. In Fig. 16 we present the assignments of this number to some of the now familiar particles.

Particle	Muon number	Particle	Muon number
π^+, π^0, π^-	0	p, n	0
μ^-	+1	e^-	0
μ^+	−1	e^+	0
ν_μ	−1	ν_e	0
$\bar{\nu}_\mu$	+1	$\bar{\nu}_e$	0

Figure 16 A table of muon numbers. Note that we have assigned different muon numbers to the neutrinos emitted with muons and the neutrinos emitted with electrons.

The first thing to notice is that we have introduced two types (families) of neutrinos—muon neutrinos and electron neutrinos. If we insist on the conservation of muon number and assign the photon muon number 0, then we forbid the decay $\mu^- \rightarrow e^- + \gamma$. If this were the only consequence of the conservation of muon number, it would not be very interesting. But we can make a prediction, namely, that the process $\bar{\nu}_\mu + p \rightarrow e^+ + n$ is absolutely forbidden, while the process $\bar{\nu}_\mu + p \rightarrow \mu^+ + n$ is allowed. This was tested in a famous experiment performed at Brookhaven National Laboratory in 1962. The muon neutrinos in this experiment came from the decay $\pi^- \rightarrow \mu^- + \bar{\nu}_\mu$. No positrons were produced when these neutrinos impinged on the target protons. The experiment clearly demonstrated that there really were two kinds of neutrinos. As we have mentioned, the muon is itself unstable. In view of the conservation of mu-ness we must then write its principle decay mode as $\mu^- \rightarrow e^- + \bar{\nu}_e + \nu_\mu$ to balance the muon numbers.

As if this were not complicated enough, in 1976 the τ lepton was discovered—a superheavy electron—with a mass energy of 1784.1 MeV. It has several decay modes involving neutrinos, and these neutrinos turned out to be distinct from both the ν_e and the ν_μ. We now had three families of neutrinos, ν_e, ν_μ, and ν_τ. The discovery of this new lepton family had been anticipated by theorists who believed in what had become known as the quark-lepton analogy. Figure 17 will make the point.

Quark		Lepton
u	$\leftrightarrow$	ν_e
d	$\leftrightarrow$	e^-
c	$\leftrightarrow$	ν_μ
s	$\leftrightarrow$	μ^-
t	$\leftrightarrow$	ν_τ
b	$\leftrightarrow$	τ^-

Figure 17 The quark-lepton analogy.

The one quark for which there is very recent experimental evidence is the t or top quark. We would have been astounded if it did not show up. In addition to the aesthetic appeal of this analogy, it appears to be needed for a consistent theory of weak interactions. Otherwise, the theory seems to be plagued with unphysical infinities. But where will it all end? How many families are there? We want to conclude this long chapter with some evidence that, at least from the point of view of the lepton families, we may have seen all of them that we are going to see. Then, in the next chapter, where we treat helium production, we will see this terrestrial experimental result confirmed cosmologically in a most spectacular way.

To begin with, we must say a few words about the notion of resonances. To make the discussion concrete we will consider processes in which very high-energy electrons and positrons are allowed to collide. For example, the Large Electron Project (LEP) at CERN in Geneva produces colliding beams of electrons and positrons, each of which has an energy of 60 GeV. At certain select energies the cross sections for these collisions are observed to rise and then fall rapidly (see Fig. 18). Such resonances can often be fit with an expression of the form

$$\sigma(E) \sim \frac{(\Gamma/2)^2}{(E - E_0)^2 + (\Gamma/2)^2} \tag{5.22}$$

Note that when $E - E_0 = \Gamma/2$, the cross section drops to half its maximum value. A way of interpreting this behavior of the cross section is to say that at the resonant energy E_0, a short-lived particle is created. The Heisenberg uncertainty principle between energy and time states that

$$\Delta E \, \Delta t \sim \hbar \tag{5.23}$$

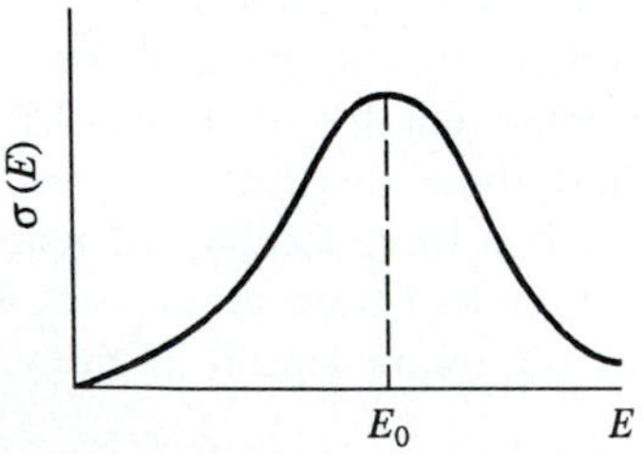

Figure 18 The resonant behavior of a cross section.

For the resonant state, this means that its mass energy is uncertain by an amount

$$\Delta m \sim \frac{\hbar}{\Delta t c^2} \tag{5.24}$$

Hence the width Γ in the resonance formula (Eq. 5.22) reflects the uncertainty in the mass of the particle due to its very short lifetime. A case in point concerns the neutral weak meson—the Z^0—which is produced in a few of the electron positron collisions in the LEP. This particle appears as a resonance in electron-positron scattering. The central energy of this resonance is given by 91.161 ± 0.031 GeV, and its width is given by $\Gamma = 2.534 \pm 0.027$ GeV. From the uncertainty principle, this corresponds to a lifetime of about 10^{-29} s. There is no way such a short lifetime could be measured directly, say, by examining the length of the track left by the particle. It can be inferred indirectly by measuring the widths of resonances.

The Z^0 has several decay modes. Each one of them is characterized by a "partial-width" Γ_i which represents the fraction of the observed width taken up by a particular decay mode. The relative fraction—the percentage—taken up by the mode i is then given by P_i, where $P_i = \Gamma_i/\Gamma$. Figure 19 lists the visible modes and their percentages. By "visible" we mean decays in which charged or strongly interacting particles are produced. Such decays are visible in a detector.

Mode	Percent
e^+, e^-	3.21
μ^+, μ^-	3.36
$\mu^\pm, e^\mp$	$<10^{-3}$
Strongly interacting	70.9
τ^+, τ^-	3.33

Figure 19 The visible decay modes of the Z^0.

The total resonant cross section for producing Z^0's is measurable. In this way one can see that visible decay modes do not account for all of the decays. They account for something over 80 percent of them. The rest of the decays are presumably of the form $Z^0 \to \nu + \bar{\nu}$, where the neutrinos can be of any flavor. The theory of weak interactions gives a very definite answer as to how much partial width each flavor of neutrino produces. The formula, which we do not derive here, is that for a single flavor f:

$$\Gamma_f = \frac{G_F M_{Z^0}^3}{12\sqrt{2}\pi} = 0.18 \text{ GeV} \tag{5.25}$$

We can decompose the total width Γ into its visible and neutrino parts:

$$1 - \frac{\Gamma_{\text{visible}}}{\Gamma} = N_f \frac{\Gamma_f}{\Gamma} \tag{5.26}$$

where N_f is the number of families. We can then use experimental results for Γ and Γ_{visible} to solve for N_f. The result is

$$N_f \simeq 3 \tag{5.27}$$

This appears to mean that the three families, e, μ, and τ, are all there are, a remarkable result which as yet has no deep explanation. Next we turn to a study of cosmological helium production where this result appears to be confirmed.

6

Helium Production

In the Micropaedia we noted that about 23 percent of the matter in the visible universe appears to be in the form of the stable isotope of helium ^{4}He. We also noted that such a large amount of helium is very difficult to account for in terms of the helium production that is continually going on in the normal course of the evolution of stars. As we saw in the introduction to this book, as far back as the late 1940s, people like George Gamow and his collaborators were already beginning to think of a Big Bang early universe explanation of the helium abundance. Their papers have a very modern flavor. The basic idea in them is to start with a very plausible initial condition and, by a series of well-understood nuclear reactions, end up with the large mass fraction of helium that astronomers see. We shall give a brief description of the way in which astronomers determine the helium abundance after we have made our theoretical exploration. Of course, we could build in an initial condition as part of which the observed helium fraction was given, but that is not playing by the rules of scientific explanation.

The helium isotope that concerns us, ^{4}He, consists of two neutrons and two protons. As we have remarked, the neutron is unstable. In 888.6 ± 3.5 s—about 15 min—the average neutron decays via the reaction $n \rightarrow p + e^- + \bar{\nu}_e$. (The neutron lifetime is not very precisely known, but this is a typical number. At the end of the chapter we shall discuss how this uncertainty affects cosmological helium production.) However, in the cosmological regime that will concern us, the reaction $\bar{\nu}_e + p \rightarrow n + e^+$ tends to replenish the neutrons which, incidentally, can also be lost by the reaction $\nu_e + n \rightarrow p + e^-$. In Chap. 4 (in the discussion preceding Eq. 4.16), we argued that until a temperature T_c determined by the relation $kT_c \simeq 10^{-3} m_p c^2$, the weak interaction, of which the above reactions are a sample, remains in equilibrium. For much of the regime of interest to us in helium production, equilibrium is maintained. We turn next to an examination of what this implies.

We argued in Chap. 4 (in the discussion preceding Eq. 4.49) that a nonrelativistic classical gas in equilibrium at a temperature T is characterized by a number density for each degree of freedom i given by

$$n_i^{\text{cl}} = e^{(\mu - mc^2)/kT} \left(\frac{mkT}{2\pi\hbar^2} \right)^{3/2} \tag{6.1}$$

Both the proton and the neutron have two degrees of freedom corresponding to spin up or spin down. Hence, in equilibrium,

$$\frac{n_n}{n_p} = \exp\left(\frac{\mu_n - \mu_p - m_n c^2 + m_p c^2}{kT}\right)\left(\frac{m_n}{m_p}\right)^{3/2} \tag{6.2}$$

We recall that

$$m_n c^2 - m_p c^2 = 1.293 \text{ MeV} \doteq Q \tag{6.3}$$

but we shall ignore this mass difference in the mass ratio in Eq. 6.2. We also recall that if the reaction $\nu_e + n \leftrightarrow e^- + p$ is in equilibrium, then the chemical potentials satisfy

$$\mu_{\nu_e} + \mu_n = \mu_e + \mu_p \tag{6.4}$$

Thus, in equilibrium,

$$\frac{n_n}{n_p} \simeq e^{(\mu_e - \mu_{\nu_e} - Q)/kT} \tag{6.5}$$

This brings us face to face with the very interesting question of the chemical potentials of the leptons.

We shall begin by discussing the chemical potentials of the electrons and positrons. In fact, in equilibrium, as particle and antiparticle, these are equal and opposite. Let us call the common magnitude of this chemical potential $|\mu_e|$. Its value is determined by making use of two general principles:

1. There does not appear to be any significant amount of antimatter within the visible universe.
2. The universe appears to be electrically neutral.

Later in the book we shall discuss the speculations that have been made about the apparent imbalance between matter and antimatter. Here we simply take it as an empirical fact. Because of charge neutrality, we must have during the regime of interest,

$$n_{e^-} - n_{e^+} = n_p \tag{6.6}$$

We remember from Chap. 4 (Eq. 4.45) that for a massless Fermi-Dirac particle with the degree of freedom i,

$$n(p)_i^{\text{FD}} = \frac{1}{e^{(pc+\mu)/kT} + 1} \tag{6.7}$$

Here μ is the chemical potential which can have either sign. Since both the electron and the positron have two degrees of freedom corresponding to their spin directions, we can write

$$n_p = n_{e^-} - n_{e^+} = 2\int \frac{d^3p}{(2\pi\hbar)^3}\left[\frac{1}{e^{(pc - |\mu_e|)/kT} + 1} - \frac{1}{e^{(pc + |\mu_e|)/kT} + 1}\right] \tag{6.8}$$

We have written this in such a way as to acknowledge the fact that $n_{e^-} > n_{e^+}$. The net charge must be negative to balance the proton charge. If we divide both sides of this equation by n_γ and remember that $\eta = n_p/n_\gamma \simeq 10^{-9}$, we conclude that $|\mu_e|/kT \ll 1$. We can make this quantitative by expanding the right-hand side of Eq. 6.8 in powers of $|\mu_e|/kT$. Keeping only the leading term and dividing by $n_\gamma = 2.4/\pi^2 \times (kT/\hbar c)^3$, we have

$$\eta = \frac{|\mu_e|}{2.4kT}\int_0^\infty dx\, x^2 \frac{e^x}{(1+e^x)^2} \simeq \frac{|\mu_e|}{kT} \tag{6.9}$$

We see, therefore, that $|\mu_e|/kT$ is negligible.

Although we believe that there are about as many cosmic neutrinos as photons, because of their weak interactions, none have so far been observed. Hence we have little or no empirical evidence about $|\mu_\nu|$, the neutrino's chemical potential. In the interest of simplicity we shall set it equal to zero. This will enable us to make some definite predictions about the helium abundance that seem to be consistent with experiment. We should keep in the back of our minds, however, that should something go wrong, we have made a definite assumption about $|\mu_\nu|$. Thus, in equilibrium,

$$\frac{n_n}{n_p} \simeq e^{-Q/kT} \tag{6.10}$$

Note that as $T \to \infty$, $e^{-Q/kT} \to 1$. This means that initially the number of neutrons and the number of protons were the same—a very agreeable initial condition.

Before we embark on the details of the theory of helium production, let us construct a little road map that reveals the highlights. The key idea is that helium is built up in the early universe by a series of nuclear fusion reactions. The triggering reaction is the radiative capture of a neutron by a proton to produce a deuteron, that is, $n + p \to d + \gamma$. However, until a critical temperature given by about 10^9 K—which corresponds to about 0.07 Mev—any deuterium formed is torn apart by the inverse reaction $\gamma + d \to p + n$. Note that this critical energy is much smaller than the deuteron binding energy of 2.23 MeV. But, in view of our discussion in Chap. 4, this should not surprise us. There we studied the same effect in the formation of neutral hydrogen by the capture process $e^- + p \to \mathrm{H} + \gamma$. Below we shall adapt that calculation to the present situation and confirm the value of the deuteron formation temperature. Once deuterons are formed, using the electromagnetic capture process, helium is rapidly built up by a sequence of strong processes of which the following are typical $d + d \to {}^3\mathrm{He} + n$, followed by ${}^3\mathrm{He} + d \to {}^4\mathrm{He} + p$. We leave it as an exercise for the reader to discover other sequences, using only the strong interactions, that produce ^{4}He starting with deuterium. There are very rapid reactions that use up essentially all of the neutrons that were available at the start of the process. The rest can end up in the trace amounts of other light nuclei such as ^{3}He, which we think also have a cosmological origin. This is a matter to which we will come back later. Hence, for all practical purposes, the ^{4}He density n_{He} is related to the neutron density n_n by the equation

$$\frac{n_n}{2} = n_{\mathrm{He}} \tag{6.11}$$

since there are two neutrons to every helium. Thus, ignoring the binding energy of helium, as compared to the rest masses, we can write (see the discussion preceding Eq. 1.49 of the Micropaedia)

$$Y \doteq \frac{M_{\mathrm{He}}}{M_{\mathrm{H}} + M_{\mathrm{He}}} \simeq \frac{4m_p n_{\mathrm{He}}}{n_{\mathrm{H}} m_p + 4m_p n_{\mathrm{He}}} = \frac{4n_{\mathrm{He}}}{n_{\mathrm{H}} + 4n_{\mathrm{He}}} \tag{6.12}$$

But we want to rewrite this in terms of the *total* proton density n_p. Since there are also two protons to every helium, we have, ignoring the protons in the other light nuclei,

$$n_p = n_{\mathrm{H}} + \frac{n_{\mathrm{He}}}{2} \tag{6.13}$$

Using Eq. 6.11 we may write

$$Y \simeq \frac{2n_n}{n_n + n_p} = \frac{2}{1 + n_p/n_n} \tag{6.14}$$

Hence the problem of calculating the helium production comes down to the problem of computing n_n/n_p. But at what temperature? It is at this point that the full complexity of the problem begins to become apparent.

As we saw in Chap. 4, the weak processes such as $\nu_e + n \leftrightarrow p + e^-$ remain in equilibrium until a temperature (see Eq. 4.16) given approximately by

$$kT_c \simeq 10^{-3} m_p c^2 \tag{6.15}$$

about 1 MeV equivalent energy. Below T_c these interactions, with one exception, become ineffective in changing neutrons into protons. The exception is the β decay of the neutron, $n \rightarrow p + e^- + \bar{\nu}_e$. This decay keeps depleting neutrons, even for $T < T_c$. Since these temperatures are below the energy equivalent of the neutron-proton mass difference of 1.293 MeV, there is not enough energy for the inverse process to take place. But it will turn out that the β decay, although it must be taken account of, has a relatively small effect on helium production. There is not that much time between the time at which the critical temperature is attained and the time at which deuteron formation takes place. Let us call this time difference t_c. We may then compute Y by evaluating Eq. 6.14 at $T = T_c$ and correcting it for β decay with a factor of e^{-t_c/τ_n}, where τ_n is the neutron's lifetime. In other words,

$$Y \simeq 2e^{-t_c/\tau_n} \frac{1}{(1 + n_p/n_n)_{T=T_c}} \tag{6.16}$$

Before we roll up our sleeves and get down to the real work of this chapter—the quantitative theory of helium production—let us finish the road map by making an extremely crude evaluation of Y by ignoring the β-decay correction and using the equilibrium value $n_p/n_n = e^{Q/kT}$. Because of the exponential character of n_p/n_n, the result will be very sensitive to the value we take for T_c. If we take $kT_c = 3$ MeV, then $n_p/n_n = 1.54$. But if we take $kT_c = 1$ MeV, then $n_p/n_n = 3.63$, which means that, crudely, Y can go from 0.79 to 0.43 when we let T_c vary by a relatively small amount. The important thing to notice is that with reasonable choices of T_c we can easily obtain a Y large enough—more than enough—to produce an affluence of cosmological helium. The main thing we have learned from our road map is that an early universe hot Big Bang explanation of the helium abundance is very plausible. We must now begin the hard work of making this explanation quantitative.

For general reasons, as well as making the β-decay correction, we need a better "clock." We need to know more accurately, during the regime of interest, the association between temperature and time. This analysis will enable us to introduce, among other things, the neutrino families of the last chapter as an element in the theory of helium production. Constructing this clock requires solving the Einstein–Friedmann equation

$$\left[\frac{\dot{R}(t)}{R(t)}\right]^2 = \frac{8\pi}{3}\rho(t)\frac{\hbar}{M_P^2 c} \tag{6.17}$$

where $\rho(t)$ is the energy density at time t. We will restrict our analysis to the "flat space" case $k = 0$. One of the difficulties in solving this problem is that $\rho(t)$ is changing, not only because the temperature on which it depends is decreasing with time but also because the particles upon which it depends disappear. For example, when $kT \simeq 1$ MeV, the electrons and positrons begin to annihilate and eventually no longer contribute to ρ. Indeed, as we shall see, after this annihilation, the clock has a different calibration. We will begin by analyzing the clock in the regime before this annihilation when the temperature is high enough so that all the relevant particles are effectively massless. The neutrons and protons are of course not effectively massless, but this is the radiation-dominated regime where they make very little contribution to the energy density. We need not make the temperature too high since then we would be including all sorts of particles that are no longer around after say 10^{-4} s which, one can show, corresponds to $kT \simeq 100$ MeV. Our clock will certainly not be accurate to 10^{-4} s.

The Einstein equation involves $R(t)$, while ρ involves $T(t)$, so we must convert one side or the other. We remind ourselves that during the normal course of the expansion of the universe, no entropy is generated. As we saw in Chap. 4 (in the discussion preceding Eq. 4.41) for essentially massless particles, the entropy density s is related to ρ by the equation

$$s = \frac{4\rho}{3T} \tag{6.18}$$

The quantity $S = R^3 s$, the total entropy, is conserved. As for essentially massless particles $\rho \sim T^4$, we conclude from the constancy of S that $R \sim 1/T$ so long as the number of particles remains constant. Hence we have

$$\frac{\dot{R}(t)}{R(t)} = -\frac{\dot{T}(t)}{T(t)} \tag{6.19}$$

This is an equation that we have made use of in Chap. 2 (see the discussion preceding Eq. 2.44) where we promised to derive it from the conservation of entropy. Thus we have

$$\frac{\dot{T}(t)}{T(t)} = -\left[\frac{8\pi}{3} \times \frac{\rho(t)\hbar}{M_P^2 c}\right]^{1/2} \tag{6.20}$$

or

$$\frac{dT}{T}\left[\frac{8\pi}{3} \times \frac{\hbar\rho(T)}{M_P^2 c}\right]^{-1/2} = -dt \tag{6.21}$$

Thus

$$t = \int_{T(t)}^{\infty} \frac{dT'}{T'}\left[\frac{3M_P^2 c}{8\pi\hbar\rho(T')}\right]^{1/2} \tag{6.22}$$

For the $k = 0$ case we are considering, this equation determines the time as a function of the temperature. One thing is apparent from the equation. If one fixes T and increases ρ by, say, adding additional particles, then for this fixed T the corresponding t is decreased. In other words, increasing the energy density causes the universe to run its course more rapidly. This, as we shall see, has essential consequences for helium production.

To proceed further, we need to know ρ. Our starting points are the expressions for $n(p)_i^{\mathrm{FD}}$ and $n(p)_i^{\mathrm{BE}}$ for effectively massless particles, namely,

$$n(p)_i^{\mathrm{FD}} = \frac{1}{e^{pc/kT} + 1} \tag{6.23a}$$

and

$$n(p)_i^{\mathrm{BE}} = \frac{1}{e^{pc/kT} - 1} \tag{6.23b}$$

We have, following our earlier discussion, set the chemical potentials equal to zero. The particles we will be dealing with are photons and leptons and, for the reasons already given, their chemical potentials can be dropped. A massless particle of momentum p has an energy pc. If we average over this energy, weighting the average by the appropriate $n(p)_i$, we can determine the average energy density ρ_i. Thus

$$\rho_i^{\mathrm{FD}} = \int \frac{d^3p}{(2\pi\hbar)^3} \frac{pc}{(e^{pc/kT} + 1)} = \frac{1}{2\pi^2} \frac{(kT)^4}{(\hbar c)^3} \int_0^{\infty} dx \frac{x^3}{e^x + 1} = \frac{7\pi^2}{240} \frac{(kT)^4}{(\hbar c)^3} \tag{6.24}$$

On the other hand,

$$\rho_i^{\mathrm{BE}} = \int \frac{d^3p}{(2\pi\hbar)^3} \frac{pc}{e^{pc/kT} - 1} = \frac{1}{2\pi^2} \frac{(kT)^4}{(\hbar c)^3} \int_0^{\infty} dx \frac{x^3}{e^x - 1} = \frac{\pi^2}{30} \frac{(kT)^4}{(\hbar c)^3} \tag{6.25}$$

Hence we arrive at the famous result that for massless particles in equilibrium at the same temperature,

$$\frac{\rho_i^{\mathrm{FD}}}{\rho_i^{\mathrm{BE}}} = \frac{7}{8} \tag{6.26}$$

We now come to an important question that we have put off discussing until we needed to, namely, How do we decide if the different particles in our cosmic soup are at the same temperature? The point is that in a certain sense equilibrium is a transitive property. If a component of a gas, say A, is in equilibrium with B, and B is in equilibrium with C, then A is in equilibrium with C. A relevant case is the one we are dealing with. Here the particles in question are electrons, photons, and neutrinos. In the regime of interest, electrons interact rapidly with photons by processes such as $e^- + \gamma \rightarrow e^- + \gamma$. These electromagnetic interactions keep the electrons and photons in equilibrium, hence at the same temperature. However, photons and neutrinos hardly interact with each other at all. We say "hardly" since some neutrino theories do endow the neutrino with very weak electromagnetic properties such as a tiny magnetic moment. This very weak interaction certainly would not keep the neutrinos in equilibrium with the photons. But in this regime the neutrinos interact rapidly with the electrons—processes like $\nu + e^- \rightarrow \nu + e^-$. Hence the neutrinos and electrons are at the same temperature. As the photons and electrons are at the same temperature, it follows that the neutrinos and the photons are at the same temperature. This argument breaks down after the electrons and positrons have annihilated since the weak interactions are not strong enough to equilibrate the neutrinos with the remaining electrons. We will return shortly to that regime. But first we want to study the temperature range $1\text{ MeV} \leqslant kT \leqslant 100\text{ MeV}$. By this epoch the mu mesons have all disappeared, but the electrons are still around. It is not quite correct to treat the electrons as massless at the lower end of the temperature range, but the error is small and the simplification great. If we included the rest mass, we would have to integrate $(p^2c^2 + m^2c^4)^{1/2}$ with a weight function of $1/(e^{\sqrt{p^2c^2+m^2c^4}/kT} + 1)$ in order to find ρ. Unlike the $m = 0$ case we cannot in the $m \neq 0$ case scale the temperature out of the integral (see the second line of Eq. 6.24), and this contribution to ρ does not have a simple power law dependence on T. That is why the massless approximation is so useful.

Suppose the number of Bose-Einstein degrees of freedom for the massless particles is N_{BE} and the corresponding number of Fermi-Dirac degrees of freedom is N_{FD}, then using Eqs. 6.25 and 6.26, we can write in this regime for ρ

$$\rho = \frac{\pi^2}{30}\frac{(kT)^4}{(\hbar c)^3}\left(N_{\text{BE}} + \frac{7}{8}N_{\text{FD}}\right) \doteq \frac{\pi^2}{30}\frac{(kT)^4}{(\hbar c)^3}N \tag{6.27}$$

where

$$N = N_{\text{BE}} + \tfrac{7}{8}N_{\text{FD}} \tag{6.28}$$

Our next task is to find N in this regime.

Since the pi and K mesons have pretty much disappeared by the time that $kT = 100\text{ MeV}$, the only bosons we have to contend with are the photons. Each of them has two degrees of freedom corresponding to the two possible polarization states. Thus $N_{\text{BE}} = 2$. There are two kinds of fermions to contend with, electrons and neutrinos and their antiparticles. Each electron has two spin states, as does each positron. On the other hand, each neutrino has one spin state corresponding to the fact that it has its spin aligned opposite its momentum. See the discussion in Chap. 5. We shall call the Standard Model, a model in which there are three, and only three, families of neutrinos: e, μ, and τ. Later we shall explore the consequences of adding additional families. Thus the standard N_{FD} in this regime is given by $N_{\mathrm{FD}} = 4 + 3 \times 2 = 10$, so that $N = \frac{43}{4} = 10.75$. Doing the integral in Eq. 6.22, we find

$$t = \frac{2.42}{(kT/\mathrm{MeV})^2}\,\mathrm{s} \times \frac{1}{\sqrt{N}} \tag{6.29}$$

In this regime $\sqrt{N} = 3.28$, so that

$$t = \frac{0.74}{(kT/\mathrm{MeV})^2}\,\mathrm{s} \simeq \frac{1}{(kT/\mathrm{MeV})^2}\,\mathrm{s} \tag{6.30}$$

The second expression is a helpful mnemonic. While this is a useful clock during the regime $1\ \mathrm{MeV} \leqslant kT \leqslant 100\ \mathrm{MeV}$, it is not the clock we need during helium production where $kT \leqslant 1$ MeV.

By the time of helium production, essentially all the positrons have disappeared in the reaction $e^+ + e^- \rightarrow \gamma + \gamma$. We may ask by what temperature has most of this disappearance taken place? We will make a very crude estimate. To reconstitute an electron-positron pair the two gammas must have at least $2m_e c^2 = 1.02$ MeV energy. We know that at every temperature there are always some gammas in the blackbody distribution with this much energy. Suppose we demand that at most one-tenth of the gamma's have an energy of $m_e c^2$ or more. At what temperature will this happen? Using the blackbody number density formula, we can write

$$\frac{\displaystyle\int_{\frac{m_e c^2}{kT}}^{\infty} \frac{x^2}{e^x - 1}\,dx}{\displaystyle\int_0^{\infty} \frac{x^2}{e^x - 1}\,dx} = \frac{1}{10} \tag{6.31}$$

If we do the integrals, we find that at $kT \simeq m_e c^2/5$ about 90 percent of the photons can no longer produce the inverse reaction $\gamma + \gamma \rightarrow e^+ + e^-$. If we use Eq. 6.30, we see that it takes about 30 s for this to take place. We will make the approximation that it happens instantaneously. In other words, we will suppose that there is a $T_>$ at which the positrons are present, followed by at $T_<$ after which they have disappeared. What has happened to the entropy? During the disappearance, the entropy of the electrons and positrons—which is approximately proportional to their number—decreases. But the number of photons has correspondingly increased. Hence the photon entropy increases to compensate for the lost electron-positron entropy. The overall entropy is conserved. But, as we saw in Chap. 4 (see the discussion following Eq. 4.66), enough electrons remain to keep the photons in a blackbody distribution throughout this epoch. Since the entropy of the photons is proportional to their number and since the number is proportional to T^3, an increase in the photon entropy means an increase in the photon temperature. The neutrinos are quite unaffected by all of this since by this time they are essentially uncoupled to both the electrons and the photons. Hence we have the relations

$$T^{\gamma}_{>} > T^{\gamma}_{<} \neq T^{\nu}_{<} \tag{6.32}$$

and

$$T^{\nu}_{>} = T^{\nu}_{<} \tag{6.33}$$

We will now make Eq. 6.32 quantitative by using the conservation of entropy.

Because of 6.33, the neutrino entropy basically drops out of the problem. But we have to satisfy

$$R^3_{<}(s^e_{<} + s^{\gamma}_{<}) = R^3_{>}(s^e_{>} + s^{\gamma}_{>}) \tag{6.34}$$

the conservation of the electron-photon total entropy. However, for massless particles we know that

$$s = \frac{4\rho}{3T} \tag{6.35}$$

(In the derivation we gave of this result in Chap. 2 we implicitly set the chemical potentials of the massless particles equal to zero. If we don't do that, the more general result is $s = (\rho + P - \mu n)/T$, where μ is the chemical potential and n is the number density. The presence of a possible chemical potential is not an issue in the argument that follows. The electrons, as we have argued, must have a very small chemical potential, and the entropy of the noncoupled neutrinos is separately conserved.) At $T^{\gamma}_{<}$ the only degrees of freedom that contribute are the photon's two polarization directions. But at $T^{\gamma}_{>}$ there are the photon's degrees of freedom along with the four electron-positron degrees of freedom. Thus, using Eqs. 6.34 and 6.35 and canceling out common numerical factors on both sides, we have

$$(R_{<} T^{\gamma}_{<})^3[2] = (R_{>} T^{\gamma}_{>})^3[2 + \tfrac{7}{8} \times 4] \tag{6.36}$$

where the numbers in the brackets are the degrees of freedom, counting the difference between the boson and fermion contributions. Thus we can conclude that

$$\left(\frac{R_{<} T^{\gamma}_{<}}{R_{>} T^{\gamma}_{>}}\right)^3 = \frac{11}{4} \tag{6.37}$$

or

$$\frac{T^{\gamma}_{<}}{T^{\gamma}_{>}} = \left(\frac{11}{4}\right)^{1/3} \frac{R_{>}}{R_{<}} \tag{6.38}$$

But the neutrino temperature always obeys $RT^{\nu} \sim$ constant. Before the neutrinos' decoupling from the electrons, they and the photon had a common temperature which obeyed $RT \sim$ constant. After the decoupling the neutrinos expand freely with the universe and continue to obey $RT^{\nu} \sim$ constant. Thus we can write

$$\frac{T^{\gamma}_{<}}{T^{\gamma}_{>}} = \left(\frac{11}{4}\right)^{1/3} \frac{T^{\nu}_{<}}{T^{\nu}_{>}} \tag{6.39}$$

But, as we have argued,

$$T^{\gamma}_{>} \simeq T^{\nu}_{>} \tag{6.40}$$

Thus we have, for all $T^{\gamma} \leqslant T^{\gamma}_{<}$,

$$T^{\gamma} = (\tfrac{11}{4})^{1/3}\, T^{\nu} \tag{6.41}$$

In other words, the cosmic background photons are hotter than the neutrinos after the electron-positron annihilation.

It may have occurred to the alert reader that this sort of thing must have gone on even earlier in the universe and, indeed, often. For example, the protons and antiprotons annihilated at kT somewhat less than 1 GeV, dumping their rest energies into the cosmic soup, to say nothing of the mu mesons and pi mesons that annihilated a little later. Why weren't the various temperatures out of whack before electron-positron annihilation? The answer is of course equilibrium. There were always enough rapidly interacting particles around to readjust the temperatures to a common value. The electron is the least massive charged lepton and, when its number was reduced, the temperature equilibrium could no longer be maintained.

Armed with this new fact about the neutrino temperature, we can return to our construction of the clock for temperatures $T^{\gamma} \leqslant T^{\gamma}_{<}$, that is, less than 1 MeV. During this regime—at least until we come to the end of radiation dominance—no massive particle contributes significantly to the energy density ρ. To find the clock we must rewrite the Einstein equation in terms of T^{ν} using the fact that $RT^{\nu} \sim$ constant. This means that we can express t in terms of the integral

$$t = \int_{T^{\nu}}^{\infty} \frac{dT^{\nu'}}{T^{\nu'}} \left[\frac{3M_P^2 c}{8\pi\hbar\rho\,(T^{\nu'})} \right]^{1/2}$$

with $\rho\,(T^{\nu})$ given by

$$\rho\,(T^{\nu}) = \frac{\pi^2}{30} \frac{(kT^{\nu})^4}{(\hbar c)^3} \left[2\left(\frac{11}{4}\right)^{4/3} + \frac{7}{8} \times 6 \right] \tag{6.42}$$

The quantity in square brackets is the effective number of degrees of freedom N. If we add up the numbers, we find $N \simeq 13.0$ or $\sqrt{N} \simeq 3.61$. If we do the integral we have, again using the relation $T^{\gamma} = (\frac{11}{4})^{1/3} T^{\nu}$ and referring to Eq. 6.29,

$$t = 2.42\left(\frac{11}{4}\right)^{2/3} \frac{1}{(kT^{\gamma}/\mathrm{MeV})^2\sqrt{N}}\,\mathrm{s} = \frac{4.75}{(kT^{\gamma}/\mathrm{MeV})^2\sqrt{N}}\,\mathrm{s} \tag{6.43}$$

or, using $\sqrt{N} = 3.61$, we have in this regime

$$t = \frac{1.32}{(kT^{\gamma}/\mathrm{MeV})^2}\,\mathrm{s} \tag{6.44}$$

We can use this clock, for example, to find the time at $kT^{\gamma} = 1$ eV—the temperature at which radiation dominance ends. This is $t \simeq 10^{12}$ s, a result that agrees with the one we found in Chap. 2 (see Eq. 2.51).

More interesting to us is the time at which deuterium can first form. We are going to go about finding it by first determining the critical temperature using arguments now familiar to us from Chap. 4 (see the discussion following Eq. 4.52). In the present case we can write

$$\frac{n_d}{n_n n_p} = \frac{g_d}{g_p g_n}\left(\frac{m_n m_p}{m_d} \times \frac{kT^\gamma}{2\pi\hbar^2}\right)^{-3/2} e^{B/kT^\gamma} \tag{6.45}$$

There are several remarks to be made about this expression. It is based on the assumption that the reaction $\gamma + d \leftrightarrow p + n$ has maintained equilibrium. This means that the chemical potentials obey

$$\mu_n + \mu_p = \mu_d \tag{6.46}$$

so they cancel out in the ratio. The B that occurs here is the binding energy of the deuteron:

$$(m_n + m_p - m_d)c^2 \doteq B = 2.23 \text{ MeV} \tag{6.47}$$

To determine the statistical weight factors, the g's, we note that the neutron and proton each have spin $\frac{1}{2}$, so each has a g of 2. On the other hand, the deuteron is a spin-1 particle, so it has a g of 3, corresponding to the three magnetic quantum numbers $+1$, 0, and -1. In the mass ratio it is reasonable to ignore the binding energy and the neutron-proton mass difference, setting $m_n m_p/m_d \simeq m_p/2$. We shall introduce the total baryon density n_B, a quantity that represents the number of baryons per unit volume, both bound and unbound. In the regime of interest the bound baryons are all in the deuterium nuclei, which have one proton and one neutron. Thus

$$n_B = n_p + n_n + 2n_d \tag{6.48a}$$

where n_p and n_n represent the free nucleon densities and n_d is the density of deuterium. Strictly speaking, we should define n_B as the net baryon density subtracting off the antibaryons. But since their number density is negligible in this epoch, we neglect them. It is convenient, and customary, to introduce the three dimensionless ratios $X_n = n_n/n_B$, $X_p = n_p/n_B$, and $X_d = n_d/n_B$. Furthermore, we can multiply and divide Eq. 6.45 by the photon number density n_γ. In addition, we shall introduce a modification of the definition of the parameter η which we have up until now defined to be simply n_p/n_γ. It is better to let

$$\eta = \frac{n_B}{n_\gamma} \tag{6.48b}$$

where n_B is the net baryon density defined above. The quantity n_B, which includes all baryons, is what occurs in these computations. This includes the neutrons which are nearly as copious as the protons during this epoch. Furthermore, the *total* baryon number $N_B = R^3 n_B$ is conserved, while $R^3 n_p$ is not, since in this regime, free protons can form nuclei. If we define a quantity G_{np} such that $G_{np} \doteq X_n X_p / X_d$, we have, putting all the numerical factors together,

$$G_{np} = 2.4\frac{\sqrt{\pi}}{6}\left(\frac{m_p c^2}{kT^\gamma}\right)^{3/2} e^{-B/kT^\gamma} \times \frac{1}{\eta} = 0.12\left(\frac{m_p c^2}{kT^\gamma}\right)^{3/2} e^{-B/kT^\gamma} \times \frac{1}{\eta} \tag{6.49}$$

As we see from its definition, G_{np} is a measure of the success of deuteron formation at a given temperature. Since $1/\eta$ is a very large number—on the order of 10^9—we see that B/kT^γ must be considerably greater than 1 before any substantial amount of deuterium can be formed. In other words, the ambient temperature must fall well below the deuteron binding energy. To make this more quantitative, let us use as our criterion of successful deuteron formation that $G_{np} = 1$. Then, if we take the logarithm of both sides of Eq. 6.49 and drop the term that goes as $(T^\gamma)^{-3/2}$, we get an equation for the temperature, call it T_c^γ, at which substantial deuteron formation takes place. Thus

$$2.10 - \ln(\eta) \simeq \frac{B}{kT_c^\gamma} \tag{6.50}$$

If we take $\eta = 10^{-9}$, we find $kT_c^\gamma \simeq B/18.63$ or $kT_c^\gamma \simeq 0.12$ MeV. Using our clock, this corresponds to a time of about $1\frac{1}{2}$ min. We should note that there is considerable uncertainty in the measured value of η. What is subject to direct measurement is the luminosity per unit volume, L, emitted by the galaxies. However, as we discussed in the Micropaedia, much of the matter in galaxies does not shine. Hence what must be done is to multiply the observed luminosity density by a so-called mass-to-light ratio M/L which measures the total amount of matter as compared to the luminous part. This is usually given in solar units, with $M_\odot/L_\odot$ being the unit. Astronomers measure M/L in a large number of galaxies in which the effect of dark matter is observable indirectly and make the correction, which is substantial. For example, in these units, the mass-to-light ratios for galaxies can be several hundred. On this basis many cosmologists take $\eta = 5 \times 10^{-10}$, but since it enters logarithmically in the equation for T_c^γ, this choice does not change the results significantly.

Now that we have an estimate for T_c^γ, we might be tempted to plug it into Eq. 6.16, using the equilibrium value $n_p/n_n = e^{Q/kT^\gamma}$, to determine Y. If we do so, we will be in for an unpleasant surprise. If we take $kT_c^\gamma = 0.12$ MeV, for example, we will find $n_p/n_n \simeq 46{,}630$, which means that for all practical purposes $Y = 0$. What has gone wrong?

The answer is equilibrium. We saw earlier that the weak interactions go out of equilibrium at $kT^\gamma \simeq 1$ MeV. But we are applying an equilibrium value for n_p/n_n for $kT^\gamma \simeq 0.10$ MeV. Since we are dealing with an exponential behavior, this has gotten us in trouble. In fact, at about $kT^\gamma = 1$ MeV, apart from β decay, the neutrons and protons have stopped interconverting because they have become decoupled from the neutrinos. The ratio n_p/n_n has become "frozen" in. This is a nonequilibrium phenomenon, and if we insist on treating the neutrons and protons as if they are still in equilibrium, we will get nonsense. The equilibrium value of the ratio n_p/n_n goes rapidly to zero as T^γ goes to infinity. Before we discuss what to do, let us make sure that none of the other calculations we have done are tainted by an unwarranted assumption of equilibrium. What about the clock? Until the electrons and positrons have annihilated, they, and the photons, are kept in equilibrium by electromagnetic interactions. This equilibrium is maintained right through to the recombination regime. So we are entitled to use the equilibrium expressions for the electrons and photons in constructing the clock. The neutrino requires a somewhat different consideration. Neutrinos do not interact after they fall out of equilibrium. In that sense they are like the photons after recombination. We saw that the photons preserve their blackbody-ness after they stop interacting and simply red-shift with the universe. This is also what happens to the neutrinos. They preserve their equilibrium spectrum with their own private temperature. Hence if we take this into account, as we have done, we are entitled to use the equilibrium expressions in constructing the clock. But what about the deuteron? In determining the temperature T_c^γ we used equilibrium values in constructing G_{np}. Was this legal? To see that it was requires a little calculation. We will start it off and leave the finishing touches for the reader. The process $n + p \rightarrow d + \gamma$ is exothermic. Energy is given off. This means that there is a nonzero rate for radiative deuteron formation even if the neutron and proton are at rest. What this means is that the quantity σv is a constant, where σ is the cross section and v is the relative velocity of the neutron and proton, or that $\sigma \sim 1/v$. We will take the theoretical value for σv for the radiative capture of a neutron by a proton to form a deuteron from a book: $\sigma v = 4.55 \times 10^{-20}\,\text{cm}^3\,\text{s}^{-1}$. To determine the rate, we need to put in the number density n_n of the neutrons. But we can write n_n approximately as $n_n \simeq \eta n_\gamma$. We now ask at what temperature this rate is equal to the expansion rate of the universe as given by our clock. Thus we solve

$$\left(\frac{kT^\gamma}{\text{MeV}}\right)^2 \text{s}^{-1} = \eta \times \frac{2.4}{\pi^2}\left(\frac{kT^\gamma}{\text{MeV}}\right)^3 \left(\frac{\text{MeV}}{\hbar c}\right)^3 4.55 \times 10^{-20}\,\text{s}^{-1} \tag{6.51}$$

The reader can solve this for kT^γ and be persuaded that we were justified in using equilibrium values in G_{np} to determine the onset of deuterium formation. Once deuterium begins to form, then we are no longer justified in using equilibrium values since deuterium is being formed precisely because the rates for $p + n \rightarrow \gamma + d$ and $\gamma + d \rightarrow p + n$ are not the same. Now to return to the helium.

The real difficulty in the theory of helium production is that, just when we would like to use it, the equilibrium approximation breaks down. We are forced to solve the actual rate equations that determine the rate at which neutrons are turned into protons by weak interactions, and vice versa. Once we understand a small wrinkle due to the expansion of the universe, it is easy enough to write down these equations. Solving them is another matter. The wrinkle is as follows. Even if all the weak interactions were turned off, we still would not have, for example, $(d/dt)n(t)_n = 0$. The reason is that the universe is expanding and the number densities are decreasing since space is blowing up. The correct equation that takes this into account is obtained by making the replacement $d/dt \rightarrow d/dt + 3\dot{R}(t)/R(t)$. We see that this is right in the limit in which the interactions are turned off since then we certainly would have $(d/dt)[R(t)^3 n_n] = 0$ which is what the replacement would tell us. But we can get around the complication caused by the expanding universe if we consider the equation for $X_n = n_n/n_B$. Note that

$$\frac{d}{dt}X_n(t) = \frac{d}{dt}\left[\frac{R(t)^3 n(t)_n}{R(t)^3 n(t)_B}\right]$$

$$= \frac{1}{R(t)^3 n(t)_B}\frac{d}{dt}\left[R(t)^3 n(t)_n\right]$$

$$= \left[\frac{d}{dt} + \frac{3\dot{R}(t)}{R(t)}\right] \times \frac{n(t)_n}{n(t)_B} \tag{6.52}$$

We have taken advantage of the fact that $R^3 n_B$ is constant in time to introduce the replacement time derivative. In the ratio X_n the expanding universe factor is automatically taken into account.

In the regime of interest—around 1 MeV—nuclei are not yet being formed in any significant numbers. Thus the baryons are free. This means that

$$X_n + X_p = 1 \tag{6.53}$$

since $n_p + n_n = n_B$. Let us call the rate at which protons are transferred into neutrons by the weak interactions $\lambda_{pn}(t)$, and the rate at which neutrons are transferred into protons $\lambda_{np}(t)$. These rates include all the processes one can imagine for making these transfers such as $\nu_e + n \rightarrow e^- + p$ and $\bar{\nu}_e + p \rightarrow e^+ + n$. In terms of these rates, we can write the rate equation for $X_n(t)$ which is the quantity we need in the helium production, $X_n = n_n/(n_n + n_p) = 1/[1 + (n_p/n_n)]$, as

$$\frac{d}{dt}X_n(t) = \lambda_{pn}(t)X_p(t) - \lambda_{np}(t)X_n(t) = \lambda_{pn}(t)[1 - X_n(t)] - \lambda_{np}(t)X_n(t) \tag{6.54}$$

This equation takes into account both the loss of neutrons and their replenishment. To get a feel for the equation, let us suppose that $\lambda_{pn}(t)$ is zero and that $\lambda_{np}(t)$ is a constant independent of time. Then we have

$$\frac{d}{dt}X_n(t) = -\lambda_{np}X_n(t) \tag{6.55}$$

which has as its solution

$$X_n(t) = X_n(0)e^{-\lambda_{np}t} \tag{6.56}$$

which is the familiar exponential decay law. We have arbitrarily called the initial time zero. The next most complicated case, and one that is closer to our situation, is to set λ_{np} and λ_{pn} equal to nonzero constants. If we solve Eq. 6.54 subject to these conditions, we find

$$X_n(t) = e^{-(\lambda_{np}+\lambda_{pn})t}X_n(0) + \frac{\lambda_{pn}}{\lambda_{np}+\lambda_{pn}}(1 - e^{-(\lambda_{np}+\lambda_{pn})t}) \tag{6.57}$$

which reduces to Eq. 6.56 in the limit in which $\lambda_{pn} = 0$. The new feature is the second term, which describes the replenishment of the neutrons. As $t \rightarrow \infty$,

$X_n(t) \to \lambda_{pn}/(\lambda_{np} + \lambda_{pn})$, which is a measure of the success of the neutron replenishment. In the case of the expanding universe, $t \sim R/\dot{R}$. When we are close to equilibrium, the λ's are greater than the expansion rate $\dot{R}/R$. This implies that the exponential terms can be neglected in Eq. 6.57, so that

$$X_n(t) \simeq \frac{1}{1 + \lambda_{np}/\lambda_{pn}} \tag{6.58}$$

But the equilibrium value of $X_n(t)$ is $1/(1 + e^{Q/kT^{\gamma}})$. Thus, in equilibrium, we have the condition that $\lambda_{np}/\lambda_{pn} = e^{Q/kT^{\gamma}}$. This relation between the rates in equilibrium is an example of what is called detailed balance. It also can be derived directly from Eq. 6.54 by invoking the equilibrium condition $d/dtX_n(t) = 0$ and using the equilibrium value for X_n. Note that the time dependence of X_n comes essentially through its temperature dependence. We can always use the entropy conservation condition $RT \sim$ constant—the T that occurs here is T^{γ} before the electron-positron annihilation and then T^{ν} afterward—to eliminate R. Thus

$$\frac{d}{dt} X_n(t) = \frac{\dot{T}(t)}{T(t)} T(t) \frac{dX_n(T)}{dT}$$

Hence a sufficient condition for approximate equilibrium is that $|\dot{T}(t)/T(t)| \ll \lambda_{pn}$, λ_{np}; that is, the rate of change of the appropriate temperature is much less than the reaction rates.

We may now give the general solution to Eq. 6.54 in terms of $\lambda_{pn}(t)$ and $\lambda_{np}(t)$. The virtue of doing this is that we can see the beginnings of the breakdown in equilibrium. We shall present the solution, and the reader can verify that it is a solution and that it includes the two special cases we have just discussed. We define a quantity $\Lambda(t)$ as

$$\Lambda(t) = \lambda_{np}(t) + \lambda_{pn}(t) \tag{6.59}$$

and a quantity $I(t, t')$ as

$$I(t, t') = \exp\left[-\int_{t'}^{t} dt''\, \Lambda(t'') \right] \tag{6.60}$$

Then the general solution to Eq. 6.54 is

$$X_n(t) = \int_0^t dt'\, I(t, t')\lambda_{pn}(t') + I(t, 0)X_n(0) \simeq \int_0^t dt'\, I(t, t')\lambda_{pn}(t') \tag{6.61}$$

We have dropped the $I(t, 0)$ term since in this quasi-equilibrium regime the rates are very large and $I(t, 0)$ is an integral over a negative exponential involving these rates. Before we indulge in the little mathematical trickery needed to reach our goal, let us say what we want to do. We are going to exploit the fact that we are close to equilibrium to drop terms that depend on anything more rapidly varying than the first derivative of the rates with respect to the time. This gives us the leading correction to the model we discussed above in which these rates were taken to be constants. It will turn out that this correction can be expressed in terms of the equilibrium form of $X_n(t)$, but at an earlier time than if the true equilibrium situation obtained. But an earlier time means a higher temperature, which means a larger $X_n(t)$ than one would have had if the situation were that of true equilibrium. In other words, if one fixes the ambient temperature of the universe at some value, say T^γ, and compares the equilibrium and the corrected X_n, the corrected X_n will be larger. It is beginning to approach the constant value it will have when the neutrinos decouple from the nucleons. By doing this calculation we will get an insight into how this process begins.

The first bit of trickery is the following identity which the reader should have no problem verifying:

$$I(t, t') = \frac{1}{\Lambda(t')} \frac{d}{dt'} I(t, t') \tag{6.62}$$

This identity can be fed directly into Eq. 6.61 and the result integrated by parts. We find

$$X_n(t) = \frac{\lambda_{np}(t)}{\Lambda(t)} - \int_0^t dt' \, I(t, t') \frac{d}{dt'} \left[\frac{\lambda_{pn}(t')}{\Lambda(t')} \right] \tag{6.63}$$

The first term in Eq. 6.63, if we use the detailed balance condition appropriate to equilibrium, is the equilibrium value of $X_n(t)$. The second term is the first correction. To make it explicit, we can once again use Eq. 6.62 and integrate by parts, keeping only the terms involving the first derivative unraised to any algebraic power. Thus

$$X_n(t) \simeq \frac{\lambda_{pn}(t)}{\Lambda(t)} - \frac{1}{\Lambda(t)} \frac{d}{dt} \frac{\lambda_{pn}(t)}{\Lambda(t)} \tag{6.64}$$

But this is just the expression we have been looking for. To see its significance let us change variables from time to temperature and use the pre-electron-positron annihilation condition that $T^\gamma R \sim \text{constant}$. To save writing, we call $\lambda_{pn}(t)/\Lambda(t) \doteq X_n(t)_{\text{eq}}$ in honor of the fact that in the Taylor series we are developing it is the equilibrium value of $X_n(t)$. Making the change of variable and using the relation between T^γ and R, we have.

$$\begin{aligned} X_n(T^\gamma) &\simeq X_n(T^\gamma)_{\text{eq}} + \frac{1}{\Lambda} \times \frac{\dot{R}(t)}{R(t)} \times T^\gamma \frac{d}{dT^\gamma} X_n(T^\gamma)_{\text{eq}} \\ &\simeq X_n \left\{ T^\gamma \left[1 + \frac{1}{\Lambda} \times \frac{\dot{R}(t)}{R(t)} \right] \right\}_{\text{eq}} \end{aligned} \tag{6.65}$$

Thus the first correction to the equilibrium distribution appears as the equilibrium distribution but is evaluated at the effective higher temperature $T^\gamma_{\text{eff}} = [1 + 1/\Lambda \times \dot{R}/R]T^\gamma$. Since $\dot{R}/R > 0$, we have $T^\gamma_{\text{eff}} > T^\gamma$, which implies that the departure from equilibrium is in the direction of increasing X_n—the first step toward its approach to a constant value.

Up to this point we have not had to know anything about the functional forms of the weak interaction rates except that $1/\Lambda \times \dot{R}/R < 1$, that is, that the reaction rates exceed the expansion rate. To proceed further, we have to introduce the functional form of these rates. We must know their explicit dependence on time in order to do the integrals in Eq. 6.61. But these time dependences are not trivial, especially if we include the effect of the electron masses. Indeed, carrying out these integrals has engaged some of our most intelligent computers. However, by making some plausible simplifying assumptions such as using the detailed balance relation among the reaction rates and taking the classical approximation to the Bose–Einstein and Fermi–Dirac distributions, one can obtain a semianalytical model in which the integrals are doable on a pocket calculator. We refer the interested reader to the reference where this is done and here content ourselves with the results. It is convenient to introduce the dimensionless variable $y = Q/kT^{\gamma}$. Figure 1 is a plot of $X_n(y)$ as computed in the model along with the equilibrium value. Note how the equilibrium value plummets to zero, while the true $X_n(y)$ "freezes". In the three-neutrino-family Standard Model the asymptotic freeze-out value, which we call $X(y = \infty)$, is given by $X(y = \infty) = 0.151$.

We wish to use this result to compute Y which is given by Eq. 6.16. We must then evaluate the β-decay correction. To do this, we have to have a more accurate determination of the time and temperature at which the fusion reactions leading to helium production take place. We have estimated this by using the G_{np} of Eq. 6.49. But this is somewhat rough since once the process begins, equilibrium is no longer maintained. To get a better answer we must go to the rate equations for deuterium production. When this is done, one finds that kT_c^{γ} is reduced from 0.12 MeV to 0.086 MeV. Using our clock, Eq. 6.44, this corresponds to a time of 179 s—about 3 min. Knowing this, and the neutron's lifetime, we are now in a position to predict Y, the helium abundance in the standard three-neutrino-family model. Thus

$$Y \simeq 2 \times e^{-179/888.6} \times 0.151 = 0.247 \tag{6.66}$$

How does this compare to the experimental number? First, a word as to how the experimental number is arrived at is in order. The problem observers have is separating the helium that is constantly being produced in stars from the primordial helium. Both give off the same spectral lines. The basic idea used is that in the course of stellar evolution, stars also produce metals such as iron. Thus it is a fair presumption that if one sees helium in a galactic object but little or no metallic contribution, then the helium is primordial. There are such objects—so-called blue

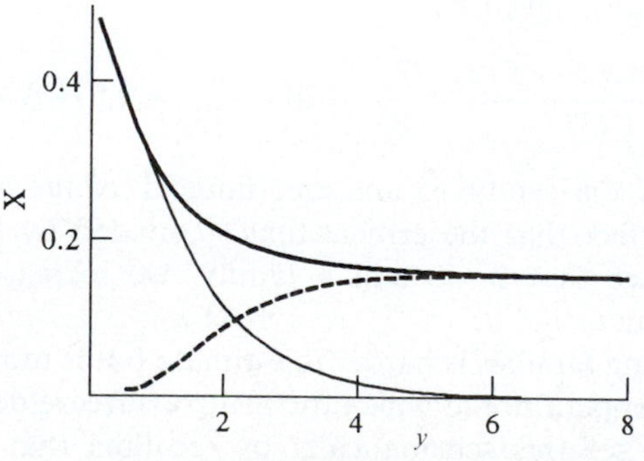

Figure 1 A plot of $X_n(y)$ in the three-family model, along with a plot of the equilibrium $X_n(y)$. Note the freeze-out.

compact galaxies being an example. By studying a variety of these, astronomers have come up with an experimental Y of 0.245 ± 0.003. We shall discuss the dependence of the theoretical Y on the various parameters presently, but for comparison we may note that computing machine calculations of Y, taking $\eta = 10^{-10}$, give in the three-neutrino-family model a value of $Y = 0.230$ with corrections. Taking these corrections into account, one can fit the data with three neutrino families. The question is what would happen if there were another family? How would that affect helium production? We shall first make a rough intuitive pass at the problem and then present the result of the simplified model we used above.

There are two sources of the change in the helium abundance as a function of a change in the number of neutrino families. Both of them are related to the same phenomenon, namely, when one adds a new neutrino family, the evolution of the universe speeds up. As Eq. 6.29 shows, $t \sim 1/\sqrt{N}$. This affects the time at which neutrons can be captured by protons to make deuterium. Since no matter what the rate of the expansion of the universe is, the capture happens at the same temperature, and so it will, according to Eq. 6.29, happen at an earlier time. This decreases the size of the β-decay contribution since the neutron lifetime does not depend on the number of families. The shorter the time available for neutron decay, the fewer the neutrons that decay into protons. If one decreases the size of the β-decay contribution, one increases the size of the helium production (see, for example, Eq. 6.66). This is an effect we can estimate rather simply and reliably. From Eq. 6.43 we see that if we fix the temperature, say at the critical temperature for forming deuterium, then

$$\frac{\Delta t_c}{t_c} = -\frac{1}{2} \times \frac{\Delta N}{N} \tag{6.67}$$

But from Eq. 6.42 we see that the N that is relevant here is given by

$$N = \tfrac{7}{4} N_\nu + 2 \times (\tfrac{11}{4})^{4/3} \tag{6.68}$$

where N_ν is the number of neutrino families—three in the Standard Model. Therefore in the three-neutrino-family model $N \simeq 13$. Let us call $Y(N_\nu)$ the helium abundance with N_ν neutrino families. Then the percentage change caused by changing the number of families by a small amount, treating $\Delta t_c/t_c$ as a small number in Eq. 6.16, and expanding the exponential is given by

$$\frac{\Delta Y(N_\nu)}{Y(N_\nu)} \doteq \frac{Y(3 + \Delta N_\nu) - Y(3)}{Y(3)} = \frac{7}{8} \times 0.20 \times \frac{\Delta N_\nu}{13} = 0.014\, \Delta N_\nu \tag{6.69}$$

where we have used both Eqs. 6.67 and 6.68 and the "fiducial" value of 13 when N_ν is 3. We have also used the fact that the critical time t_c divided by the neutron's lifetime is about 0.20. We see that if we add a family, we increase the helium production by over 1 percent.

The second effect of adding families is harder to estimate but is more significant. The general idea is that the temperature at which the neutrons freeze depends on the number of families. We can see this schematically by recalling that the neutrino contribution to $\dot{R}/R$ is proportional to $\sqrt{2N_\nu}T^2$, while the weak interaction rate is proportional to T^5. The factor of 2 is because we must count the neutrino and antineutrino for each family type. Hence if we equate these rate expressions, we see that the freeze-out temperature is proportional to $N_\nu^{1/6}$. Thus, qualitatively, we see that the more families there are, the higher the freeze-out temperature. This means that there will be more helium produced when we add additional families. The difficulty is estimating how much more. It is, as we will see, easy to estimate this in the equilibrium approximation. To do better, we must either resort to computing machines or to the simplified model. We begin by using the equilibrium approximation that will give us the orders of magnitude. We call $X_n(N_\nu)$ the ratio $1/(1+n_p/n_n)$ evaluated for N_ν neutrino families. We may, assuming that the family correction is small, expand $X_n(N_\nu)$ around $N_\nu = 3$. Thus

$$X_n(N_\nu) \simeq X_n(3) + \frac{\Delta X_n(N_\nu)}{\Delta N_\nu}\Big|_{N_\nu=3}\Delta N_\nu \tag{6.70}$$

But from the definition,

$$\frac{\Delta X_n(N_\nu)}{\Delta N_\nu} = -X_n(N_\nu)^2\frac{\Delta n_p/n_n}{\Delta N_\nu}\Big|_{N_\nu=3} \tag{6.71}$$

It is at this point that we introduce the equilibrium value for n_p/n_n. We do this for a temperature corresponding to about 1 MeV—the freezing temperature where the equilibrium approximation is breaking down. That is why we do not claim high precision for the result, although we do expect it to be about the right order of magnitude. If we call the freezing temperature T_c, not to be confused with T_c^γ, the temperature at which the deuterons began to form, we have from our previous discussion the relation

$$\frac{\Delta T_c}{T_c} = \frac{1}{6}\frac{\Delta N_\nu}{N_\nu} \tag{6.72}$$

Taking $n_p/n_n = e^{Q/kT_c}$, we have

$$\frac{\Delta(n_p/n_n)}{\Delta N_{\nu c}} = -\frac{1}{6}\times\frac{Q}{kT_c}\times\frac{n_p/n_n}{N_\nu} \tag{6.73}$$

where we have used Eq. 6.72 to evaluate $\Delta T_c/\Delta N_\nu$. Thus we have

$$\Delta X_n(N_\nu) = X_n(N_\nu) - X_n(3) = X_n(3)^2\frac{Q}{kT_c}\times\frac{\Delta N_\nu}{6} \tag{6.74}$$

We can evaluate this for, say $T_c = 1$ MeV—the result is essentially the same if we take 3 MeV—and find $\Delta X_n(N_\nu) \simeq 0.01\,\Delta N_\nu$. Since the helium yield is proportional to $2X_n(N_\nu)$, in the equilibrium approximation the change in the yield is about 2 percent per family. This, it turns out, underestimates the result found by more careful computations. If we use the nonequilibrium model, we get $\Delta Y(N_\nu)/Y(N_\nu) = 0.042\,\Delta N_\nu$ from the change in the freeze-out temperature. Hence the combined change in the helium yield from both of these effects of changing the number of families is given by $\Delta Y(N_\nu)/Y(N_\nu) = 0.056\,\Delta N_\nu$, which is nearly 6 percent! But experimenters are now claiming an accuracy of ± 0.003 in their measurement of Y. They assert therefore that a model with four neutrino families is essentially ruled out, while three is in comfortable agreement with the data. But this is the result that the people who measured the width of the Z^0 found! Only three neutrino families were compatible with the Z^0-width experiments. Before we comment on this remarkable concordance we must finish our discussion of how the helium production varies as a function of the rest of the parameters.

Up to this point, we have treated the neutron lifetime τ_n as if it were as well known as the speed of light. It isn't. It is a very difficult quantity to measure. To give some idea of progress, the 1956 monograph of Hans Bethe and Philip Morrison—*Elementary Nuclear Theory*—quotes a value of τ_n of 1111 ± 140 s, while Weinberg in his 1972 book *Gravitation and Cosmology* uses a value of 1,013 s. We have taken our value of 888.6 ± 3.5 s from the 1990 Particle Properties Data Booklet, although the 1992 version of the same gives 889.1 ± 2.1 s, with a fairly wide scatter from the different experiments. It is evident that we need to take into account possible future changes in the measured τ_n. The first effect of a change in τ_n is obvious enough, the exponential in the decay law changes: $e^{-t_c/\tau_n} \rightarrow e^{-t_c/(\tau_n + \Delta\tau_n)}$. If $\Delta\tau_n/\tau_n \ll 1$, we can expand the denominator and the exponential to learn that the percentage change in Y from this source is given by $t_c/\tau_n \times \Delta\tau_n/\tau_n \simeq 0.20\,\Delta\tau_n/\tau_n$. However, there is a more substantial, and more difficult to calculate, change in Y that occurs whenever the neutron lifetime is remeasured. The weak coupling constant G_F is determined by measuring the neutron's lifetime. Indeed, when one actually calculates the rates λ_{pn} and λ_{np}, it is convenient to use the neutron's lifetime to set the scale of these rates. If the lifetime turns out to be longer than the presently accepted fiduciary value, it means that the rates are somewhat smaller than we think. This means that the freeze-out will occur earlier since the expansion rate will exceed the weak rates at a higher temperature. But this means that more helium will be produced. To find out how much more requires a calculation using the actual rates. The equilibrium approximation won't help us since the specific values of the rates do not enter the equilibrium value for Y. If we use the nonequilibrium model, we find that this contribution to the percentage change in Y is given by $0.52\,\Delta\tau_n/\tau_n$. Thus the entire percentage change in Y coming from a change in τ_n is given by $\Delta Y(\tau_n)/Y(\tau_n) = 0.72\,\Delta\tau_n/\tau_n$. This is potentially a large change if for some reason the experimenters find some significant error in their previous measurements. If the error is really 2 s in 889, we don't have much to worry about.

The ratio $\eta = n_B/n_\gamma$ is not that well known. We have taken for our fiduciary value $\eta_0 = 5 \times 10^{-9}$. But this could be wrong by a factor of 2. Fortunately, the effect on the uncertainty in Y is only logarithmic, as we will now show. The quantity η enters our story in the determination of T_c^γ. Recall Eq. 6.50 which gives the approximate relationship $\ln(\eta) - 2.10 \simeq B/kT_c^\gamma$. If we introduce the fiduciary η_0 and the fiduciary temperature $T_c^{\gamma 0}$, we have

$$\ln\left(\frac{\eta}{\eta_0}\right) = \frac{B}{k}\left(\frac{1}{T_c^\gamma} - \frac{1}{T_c^{\gamma 0}}\right) \doteq \frac{B}{kT_c^{\gamma 0}}\frac{\Delta T_c^\gamma}{T_c^\gamma} \tag{6.75}$$

But what do we need to know to determine $\Delta Y(\eta)/Y(\eta)$? We need the change in the time t_c corresponding to the changed T_c^{γ}, since this is what enters the β-decay term e^{-t_c/τ_n}. Clearly, then

$$\frac{\Delta Y(\eta)}{Y(\eta)} = -\frac{\Delta t_c}{t_c} \times \frac{t_c}{\tau_n} \tag{6.76}$$

where, to first order in Δt_c, we can use the fiducial values to set $t_c/\tau_n = 0.20$. But as our clock shows, $t_c \sim 1/(T_c^{\gamma})^2$. Thus

$$\frac{\Delta t_c}{t_c} = -2\frac{\Delta T_c^{\gamma}}{T_c^{\gamma}} = -2\frac{kT_c^{\gamma}}{B} \times \ln\left(\frac{\eta}{\eta_0}\right) \tag{6.77}$$

Hence

$$\frac{\Delta Y(\eta)}{Y(\eta)} = 2\frac{kT_c^{\gamma}}{B}\ln\left(\frac{\eta}{\eta_0}\right)\frac{t_c}{\tau_n} \tag{6.78}$$

With our numbers this gives $\Delta Y(\eta)/Y(\eta) \simeq 0.02 \ln(\eta/\eta_0)$. If we use the simplified model, we find $\Delta Y(\eta)/Y(\eta) = 0.036 \ln(\eta/\eta_0)$. We may now put the three corrections together and write, for the simplified model,

$$\frac{\Delta Y}{Y} = \frac{\Delta Y(N_\nu)}{Y(N_\nu)} + \frac{\Delta Y(\tau_n)}{Y(\tau_n)} + \frac{\Delta Y(\eta)}{Y(\eta)} = 0.056\,\Delta N_\nu + 0.72\frac{\Delta\tau_n}{\tau_n} + 0.036\ln\left(\frac{\eta}{\eta_0}\right) \tag{6.79}$$

If we use the value $Y = 0.247$ taken from the nonequilibrium model with three families, we can write the corrected Y in the form

$$Y = 0.247 + 0.014\,\Delta N_\nu + 0.18\frac{\Delta\tau_n}{\tau_n} + 0.009\ln\left(\frac{\eta}{\eta_0}\right) \tag{6.80}$$

It is interesting to compare this expression to the one arrived at as a fit to the sophisticated computing machine results, which we call Y_{comp}. As the reader will notice, the fiducial values used are somewhat different than ours. Thus

$$Y_{\text{comp}} = 0.230 + 0.013(N_\nu - 3) + 0.014(\tau_n^{1/2} - 10.6\text{ min}) + 0.011\ln(\eta \times 10^{10}) \tag{6.81}$$

In this expression, the neutron's half-life in minutes, $\tau_n^{1/2}$, is used, while in our expression τ_n stands for the 888.6 s mean life. In the nonequilibrium model an η_0 of 5×10^{-10} was used. In Fig. 2 we present a computer-generated graph of Y as a function of N_ν and η. We also show the experimental point. The reader will no doubt find the conclusion that $N_\nu = 3$ pretty persuasive.

What conclusions can we draw? On the face of it, we should be overjoyed at the concordance of the three neutrino families as measured in helium production and in the Z^0 decay. However, on further reflection, one's joy may be a bit tempered. Is it really true that the τ lepton, with its rest energy of 1784 MeV, is the highest-mass lepton we are ever going to discover? This certainly seems very odd, especially since we have no theoretical understanding of why the number of families is limited to three. If this is true, and the quark-lepton analogy continues to hold, then once the top quark is assured, we will have discovered all the quarks there are as well. This also seems odd. But perhaps we will be in for a surprise. In view of what we know,

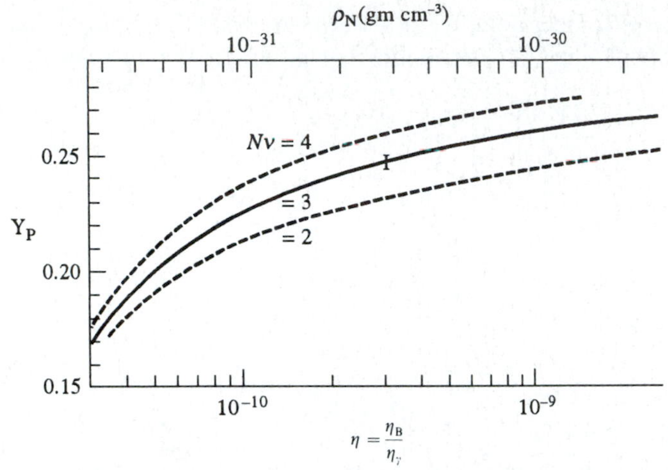

Figure 2 Y as a function of η and N_ν. Note the experimental point.

what kind of surprises can we tolerate? In order not to play tennis without a net, we will spell out our rules. For purposes of this discussion, we shall call a particle a lepton if it has weak interactions of about the weak interaction strength we are accustomed to and no strong interactions. In particular, our nominal leptons must interact with the Z^0 with the conventional strength. The only way leptonic neutrinos of this ilk could have escaped detection—since all of them contribute to the width—is if they are too massive to have been allowed to take part in Z^0 decay. Since the Z^0 has a rest energy of about 91 GeV, such hypothetical objects must have rest energies greater than about 45 GeV if they are not to be emitted in the decay $Z^0 \rightarrow \nu + \bar{\nu}$. But would such a very massive neutrino wreak havoc on the helium results? We defer a full discussion of massive neutrinos until the next chapter, but in principle it is possible to arrange things so as not to interfere with what we know about helium. Hence it is not out of the question for a generation of superheavy leptons to show up. Perhaps they belong to novel families. Time will tell.

Until the helium and the Z^0 results became firm, there was the distinct possibility that these two types of experiments might reveal different numbers of families of essentially zero-mass neutrinos. For example, the Z^0 decay might have shown more families than the helium production. Hence some thought was given as to how to reconcile such a hypothetical result. As we mentioned in the discussion following Eq. 6.9, we have no direct information about the neutrino's chemical potential. If we allow the neutrino equilibrium distribution for a given family to change from $1/(e^{pc/kT} + 1)$ to $1/(e^{(pc+\mu)/kT} + 1)$ by introducing a chemical potential, this will affect the helium production. Before we explain how, note that if there is a neutrino chemical potential, there will be a net neutrino-antineutrino number density which we will call n_L. As we saw in the discussion leading to Eq. 6.9, we can write

$$n_L = n_\nu - n_{\bar{\nu}} = \int \frac{d^3p}{(2\pi\hbar)^3} \left[\frac{1}{(e^{(pc-|\mu|)/kT} + 1)} - \frac{1}{(e^{(pc+|\mu|)/kT} + 1)} \right] \tag{6.82}$$

If we expand in $|\mu|/kT$ and divide by n_γ, as well as multiplying and dividing by n_B, we wind up with the statement that

$$\frac{n_L}{n_B} \simeq \frac{|\mu|}{kT} \times \frac{1}{\eta} \tag{6.83}$$

This is an interesting result because it tells us, given the tiny magnitude of η, that even with a very small $|\mu|$ we can still have a very large ratio of baryon to lepton asymmetry. In other words, $|\mu|$ is not much constrained by experiment. We can get some idea of how the inclusion of such a chemical potential will affect helium production by studying the equilibrium case. Referring to Eq. 6.5 for the case of a nonzero positive chemical potential for the electron neutrino, we can write for the equilibrium value of n_p/n_n,

$$\frac{n_p}{n_n} \simeq e^{(\mu_{\nu_e} + Q)/kT} \tag{6.84}$$

Thus the modified version of the equilibrium value of X_n is given by $1/(1 + e^{(\mu_{\nu_e} + Q)/kT}) \simeq e^{-(\mu_{\nu_e} + Q)/kT}$ when $e^{Q/kT} \gg 1$. The effect of a positive electron neutrino chemical potential is to reduce the helium abundance. This is also borne out in the nonequilibrium model calculation where one finds $\Delta\mu_{\nu_e} = -0.25\,\mu_{\nu_e}/kT$. Hence we could easily compensate for a new family by introducing a chemical potential. But since the data appear well-fitted by three families, we conclude that μ_{ν_e}/kT must be less than about $\frac{1}{10}$. We might as well set it equal to zero until there is a reason not to.

Before concluding this long chapter there are three things we wish to do. First, we will present a "calendar" of the events in the early universe we have discussed up to this point. So many critical temperatures and times have been mentioned that the reader might welcome a summary. Then we wish to discuss briefly the putative cosmological origins of other light nuclei which have been observed in trace amounts. Then, finally, we wish to present an argument that was first introduced by George Gamow and his collaborators Ralph Alpher and Robert Herman in the late 1940s

CALENDAR OF EARLY UNIVERSE EVENTS

kT^γ	What Happens	Particles Available
100 MeV $\leftrightarrow \sim 10^{-4}$ s	Muons begin to annihilate.	ν_e, ν_μ, ν_τ, e^-, γ, *anti'* s some p, n
2 MeV $\leftrightarrow \sim 0.2$ s	Neutrinos decouple, and n_p/n_n is frozen apart from β decay.	Same particles
1 MeV $\leftrightarrow \sim 1$ s	e^+, e^- begin to annihilate, leaving some e^-; $T^\nu \neq T^\gamma$	ν_e, ν_μ, ν_τ, γ, p, n, e^- with $n_p = n_{e^-}$
0.086 MeV $\leftrightarrow \sim 3$ min	Deuterons are formed, and helium is produced.	Same particles
1 eV $\sim 10^{12}$ s	Recombination; $e^- + p \rightarrow \mathrm{H} + \gamma$.	Same particles

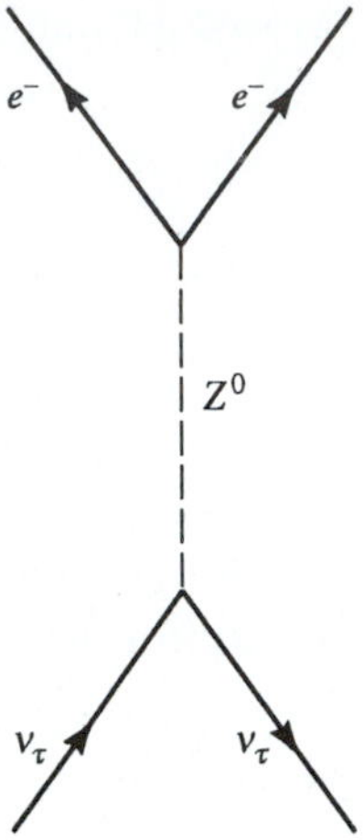

Figure 3a A τ neutrino interacting with an electron.

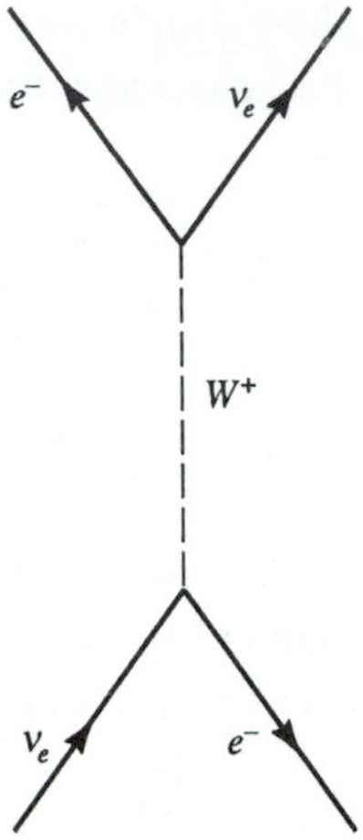

Figure 3b An electron neutrino interacting with an electron.

to predict the temperature of the cosmic background radiation. It too depends on the theory of helium production in a most interesting way.

The calendar gives us an opportunity to comment on a subtlety connected with neutrino decoupling. In this respect not all the neutrinos are created quite equally. The τ and the μ neutrinos can interact with electrons only through the exchange of a Z^0 (see Fig. 3a), while the electron neutrino can also interact with electrons with an exchange of a W meson (see Fig. 3b). This means in effect that the τ and μ neutrinos have somewhat weaker interactions with electrons than does the electron neutrino. One consequence of this is that the τ and μ neutrinos freeze out slightly earlier than the electron neutrino. This has a small effect on helium production, and we have ignored it.

Now to other light elements. The first thing we want to make clear is that in dealing with the abundances of light elements, the equilibrium approximation is not going to help us at all. We have seen this already, but we would like to make the point in a somewhat different way. It will be recalled that we defined a quantity G_{np} as the ratio $X_p X_n / X_d$, where the X's are the dimensionless ratios of the indicated particle densities to the total baryon density. We used the equilibrium value of G_{np} to estimate the critical temperature at which deuterium is formed. Let us try the same thing for ^{4}He. Since ^{4}He is produced, in essence, by a series of reactions in which two neutrons and two protons become one helium nucleus, we have, if these reactions take place under equilibrium conditions, the relation involving the chemical potentials

$$2\mu_n + 2\mu_p = \mu_{^4\mathrm{He}} \qquad (6.85)$$

This motivates us to define H_{np} as $X_n^2 X_p^2 / X_{^4\text{He}}$. In this way, the chemical potentials will cancel out in the equilibrium evaluation. By a straightforward computation, which we leave for the reader to carry out, we can show that

$$H_{np} = \frac{g_p^2 g_n^2}{g_{^4\text{He}}} \times \frac{1}{\eta^3} \times \frac{\pi^{3/2}}{2^{15/2}} \times \left(\frac{m_p c^2}{kT} \right)^{9/2} e^{-B(^4\text{He})/kT} \tag{6.86}$$

where we have, in the factor involving the proton mass, neglected the neutron-proton mass difference and the helium binding energy which we have called $B\,(^4\text{He}) = 28.30$ MeV. The ^{4}He nucleus has spin 0, which means that $g_{^4\text{He}} = 1$. Taking, as before, $\eta = 10^{-9}$ and putting $H_{np} = 1$ we can, using the same approximations as the ones we used when discussing G_{np}, find the critical temperature T_c at which helium is formed. This turns out to be given by $kT_c = 0.47$ MeV, which is about four times higher than the corresponding temperature for the formation of deuterium! This implies that helium forms before deuterium. But this is not at all what happens. The equilibrium approximation has led us astray. In reality, first deuterium is formed in the reaction $n + p \rightarrow d + \gamma$, and then helium is built up by a series of two-body reactions that have to be examined in detail using the rate equations. This means that making the correct prediction on the basis of the Standard Model for the production of such light nuclei as ^{3}He and ^{7}Li (the common isotope of lithium) requires a nontrivial calculation. We shall discuss the situation for cosmological deuterium, ^{3}He, and cosmological lithium.

Most, but not all, of the cosmological deuterium is used up making helium. Any stray deuterium one now finds in meteorites, or in the local interstellar medium, can very likely be assumed to be cosmological. This is because the deuterium made in stars will be consumed in the fusion processes such as $d + d \rightarrow {}^3\text{He} + n$ that eventually lead to the production of ^{4}He. Measurements show a ratio of deuterium to hydrogen densities of about 2×10^{-5}. Since deuterium is so easily destroyed, this is probably a lower bound on the amount of cosmological deuterium that was formed and not consumed. The formation of cosmological deuterium is a sensitive function of η. As η increases, more and more nucleons are available around the time of helium formation to convert deuterium into helium. This means that fewer deuterons are unconverted. In other words, increasing η decreases the ratio of presently observed deuterium to hydrogen. If this ratio is taken to be at least 1×10^{-5}, then it turns out that η must be less than 10^{-9}. We may, however, use a connection between η and Ω_M to tell us something about Ω_M. The connection follows from Eq. 2.20 which states that $\Omega_M = m_p \times n_B / (1.879 \times 10^{-29} \times h^2\,\text{g cm}^{-3})$. By multiplying and dividing by n_γ we can write this as

$$\Omega_M = \eta \times n_\gamma \times m_p \times \frac{10^{29}}{1.879 h^2\,\text{g cm}^{-3}} \tag{6.87}$$

If we use $n_\gamma = 417\,\text{cm}^{-3}$, we have the useful relation

$$\Omega_M = 3.76 \times 10^7 \eta \times h^{-2} \tag{6.88}$$

If we take $h > 0.4$ and $\eta < 5 \times 10^{-10}$, the presently accepted upper bound, we have $\Omega_M < 0.12$. This is a strong argument that baryonic matter cannot close the universe, i.e., make $\Omega = 1$. In a certain sense, deuterium and ^{3}He should be lumped together since the reaction $d + d \rightarrow {}^3\text{He} + n$ so readily converts deuterium into ^{3}He. There is also a corresponding reaction that makes ^{3}H—tritium. But tritium is unstable against β decay with a half-life of 12.3 yr. The combined ratio of deuterium plus ^{3}He to hydrogen is measured to be less than about 8×10^{-5} which, according to the calculations, implies that η is greater than about 3×10^{-10}. This implies a lower bound on Ω_M, namely, $\Omega_M > 0.011$. The nucleus ^{7}Li can be produced in the relatively slow electromagnetic interaction ${}^4\text{He} + {}^3\text{H} \rightarrow {}^7\text{Li} + \gamma$ or in the electromagnetic weak chain starting with ${}^4\text{He} + {}^3\text{He} \rightarrow {}^7\text{Be} + \gamma$ followed by the weak reaction ${}^7\text{Be} + e^- \rightarrow {}^7\text{Li} + \nu_e$. The amount of ^{7}Li that is produced this way is miniscule. The ratio of ^{7}Li to hydrogen is predicted to be about 10^{-10}. The claim is that such an abundance of ^{7}Li has been observed in metal-poor old stars. If this is really primordial lithium, we have another pleasing confirmation of the hot Big Bang origin of the light elements. Next, and finally, we turn to the arguments of Gamow and his collaborators.

The discussion we are going to present is an amalgam of Gamow's first paper on this subject, published in 1948, where the essential ideas were presented, although the cosmic radiation temperature had not yet been predicted, and the much longer and more careful paper, published the same year, by Alpher and Herman. This article contained the remarkable prediction the cosmic background radiation, which was not observed until 1965, should have a temperature of about 5 K. Since we are not going to be as careful as Alpher and Herman were, we will not get quite their answer. What would it take to predict the present temperature of the background radiation? Since we know that the present baryon density, which we call n_{B0}, is about $3 \times 10^{-7}\ \text{cm}^{-3}$ and that the baryon density varies as $(T^\gamma)^3$, we could find the present radiation temperature T_0^γ if we knew the baryon density n_B at any other temperature because we could use then the formula

$$(T_0^\gamma)^3 = \frac{n_{B0}}{n_B}(T^\gamma)^3 \tag{6.89}$$

to find T_0^γ. But how can we find the baryon density at an earlier temperature?

This is where Gamow's genius came in. He was committed to the idea that elements like helium had a hot Big Bang origin. He realized that the key process that started the building chain was deuteron formation—the process $n + p \rightarrow d + \gamma$ that we have been considering. He argued that this process would be fully operative when the capture rate equaled the expansion rate, i.e., when

$$n_B \sigma v \simeq \frac{\dot{R}}{R} \tag{6.90}$$

Thus

$$n_B \simeq \frac{\dot{R}}{R} \times \frac{1}{\sigma v} \tag{6.91}$$

The quantity σv one gets from theory. We used $\sigma v = 4.55 \times 10^{-20}\ \text{cm}^3\ \text{s}^{-1}$. To find $\dot{R}/R$ one needs a clock, i.e., our equation 6.30. To use the clock one must know the temperature at which the capture reaction produces deuterium. By studying the rate equations, Alpher and Herman arrived at a critical temperature of 0.6×10^9 K which corresponds to 5.17×10^{-2} MeV. We used a value of 8.60×10^{-2} MeV. If we use Alpher and Herman's number, we find that $n_B \simeq 5.9 \times 10^{16}\ \text{cm}^{-3}$. If we put these numbers into Eq. 6.89, we find that $T_0^\gamma \simeq 10$ K. Doing the more careful calculation, Alpher and Herman found 5 K. It was an astonishing prediction, and still more astonishing is the fact that so few people understood its significance. Indeed, it played no role in the actual discovery of the cosmic background radiation. In the next chapter we turn to the subject of massive neutrinos.

7

Massive Neutrinos

Pauli first introduced the neutrino hypothesis in December 1930 in a famous unpublished letter. He accepted as self-evident that the neutrino—which he called the "neutron"—had a mass. He wrote, "The mass of the neutrons should be of the same order of magnitude as the electron mass and in any case not larger than 0.01 times the proton mass." [It is interesting to note that what we call the neutron was not identified until 2 yr later by James Chadwick. Up until that time the neutral component of nuclear matter was taken to be a bound proton and electron. In fact, this is what Chadwick thought he had found.] Indeed, when in 1934 Fermi published the first theory of beta decay, he allowed for the possibility that what he then named the "neutrino" had a mass. Before we explain how Fermi proposed to determine whether or not the neutrino did have a mass, it is worthwhile to make a general remark. When a particle is massless—such as the photon—there should be a good reason. For the photon there is a good reason—a symmetry known as gauge invariance. The physical results of electrodynamics remain the same when one replaces the electromagnetic field A_μ by $A_u + \partial_\mu \Lambda$, where Λ is an arbitrary function of space and time. The choice of Λ fixes the "gauge". One can show that this symmetry is necessary for the masslessness of the photon. For the neutrino, no one has found a good reason for it to be massless, unless the fact that if it is massless it can be represented by the elegant theory in which its spin is exactly pointing opposite its momentum can be considered a good reason. Back to Fermi's paper.

In his paper, Fermi made the first computation of the beta-decay spectrum—the relative number of electrons emitted at a given kinetic energy. This spectrum is sensitive to the mass of the neutrino in two important ways. First, its upper kinetic energy T—the so-called end point—depends on the neutrino mass. Suppose the decay is of the form $A \to B + e + \nu$, where we are not being fussy about whether we are dealing with electrons or positrons or neutrinos or antineutrinos. A and B are the heavy parent and daughter nuclei, respectively. We leave it as an exercise for the reader to show that the upper end-point $T_{\max}$ is given by

$$T_{\max} = \frac{[m_A^2 + m_e^2 - (m_B + m_\nu)^2]c^2}{2m_A} - m_e c^2 \tag{7.1}$$

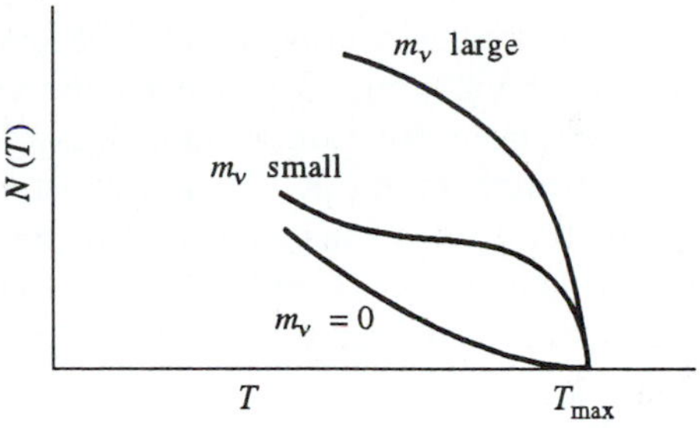

Figure 1 Fermi's graph for the beta-decay spectrum near the end point. We have translated Fermi's notation from the original German.

This follows simply from the conservation of energy and momentum. What does not follow from energy-momentum conservation, but rather from the quantum dynamics of the weak interactions that cause beta decay, is the *shape* of the spectrum. This is what Fermi computed. Figure 1 is his original graph of the spectrum near the end point where it is especially sensitive to the mass of the neutrino. The reason why it is so sensitive at the end point is that when the electron's energy is maximum, the neutrino's energy is minimum. If it has a mass, this will manifest itself in its energy when it is moving relatively slowly.

Looking at Eq. 7.1 and the graph, one might naively think that it would be a very simple matter to determine the neutrino's mass—if any. For example, if we ignore terms involving the squares of the neutrino and electron masses divided by the squares of the nuclear masses, which physically means neglecting the recoil kinetic energy of the daughter nucleus, we can use Eq. 7.1 to find a very simple expression for the neutrino mass in terms of the end point energy and m_A, m_B, that is,

$$m_\nu c^2 = (m_A - m_B)c^2 - T_{\max} \qquad (7.2)$$

The experimental problem is apparent from Fermi's graph. At the end point, where the plot is sensitive to the neutrino mass, the number of events goes rapidly to zero. In fact, if we work out the beta-decay formula, there is an expression in the spectrum that goes as $[(T_{\max} + m_\nu c^2 - T)^2 - m_\nu^2 c^4]^{1/2}$ which determines its shape near the end point. One can convince oneself that its slope goes to infinity as $T \to T_{\max}$. The spectrum goes to zero rapidly indeed! With the small number of events near the end point, it has turned out to be extremely difficult to measure the spectrum there or even to determine the end-point energy with great accuracy. To complicate matters further, the nuclei in question are in practice not bare nuclei but parts of atoms or molecules with their attendant electrons. There are, in general, complicated interactions between the beta-decay electrons and the atomic electrons, and these distort the spectra. It is therefore desirable to use a light nucleus which has very few atomic electrons and one with as small a $T_{\max}$ as possible. By plotting the expression above, one can persuade oneself that the larger $m_\nu c^2/T_{\max}$ is, the more the sensitivity of the spectrum to the neutrino mass is enhanced. This is plausible because the particles are more and more nonrelativistic. For both these reasons, it turns out that the decay $^3H \to {}^3He + e^- + \bar{\nu}_e$—the beta decay of tritium—is the one that has been most closely studied. This system is sufficiently simple that the electron interactions can be computed, and the maximum electron kinetic energy is only 18.6 keV as opposed to more typical beta decays in which the maximum electron kinetic energy is several megaelectronvolts. Experiments have been done on tritium decay for many years, including a well-known 1972 measurement by K. E. Bergkvist from which he concluded that the mass energy of the electron neutrino had to be less than 60 eV and was consistent with zero. This suited most people just fine since the standard theory, with its polarized neutrino, seemed so elegant. Then came the intimations that all might not be quite so rosy.

There were two sorts of intimations, one from the heavens—so to speak—and one from Russia. In the early 1980s a group from the former Soviet Union reported on a series of tritium experiments which purported to produce a mass energy of the electron neutrino of 30 eV with an error lying between $+2$ and -8 eV. This result stunned the cosmology community. In the meantime another enterprise had been unfolding for several years in the Homestake gold mine in South Dakota. Ray Davis and his collaborators had by 1970 installed a huge vat filled with 3.8×10^5 liters of C_2Cl_4—perchloroethylene cleaning fluid—in the mine. Why? The reason for using the mine is to provide shielding. The vat of cleaning fluid acts as a neutrino detector for neutrinos emitted by the sun. To keep from being overwhelmed by other events, one takes advantage of the fact that the neutrino—a chargeless, weakly interacting particle—can penetrate light-years of lead, to say nothing of the earth, barely interacting. Cosmic rays, which can overwhelm the neutrino events one is looking for, will be in large measure blocked by the shielding of the earth. To understand the cleaning fluid, we have to return to the mechanism that generates energy in the sun. We touched on it in the Micropaedia, but now we must go into somewhat more detail.

In the Micropaedia we mentioned that the initiating reaction in the sun is $p + p \rightarrow d + e^+ + \nu_e + 0.42$ MeV. The 0.42 MeV is the maximum kinetic energy the neutrino can have under the assumption that whatever the neutrino's rest mass is, it can be neglected compared to the other masses in the reaction. In particular, we neglect it compared to the electron's rest energy of 0.51 MeV. Given the Bergkvist upper limit on $m_{\nu_e}c^2$ of 60 eV, this is a good approximation. By this point, the alert reader may wonder why the proton-proton reaction is the trigger reaction in the sun and other stars comparable to it, while it played no role in element formation in the early universe and, conversely, why the reaction that triggered element formation in the early universe, $n + p \rightarrow d + \gamma$, plays no role in the sun. Here are some of the principal differences. There are no free neutrons in the sun because of the neutron's instability, while at the time of element formation in the early universe, the ratio of neutrons to protons was about 1:3. Therefore the sun cannot take advantage of the comparatively rapid electromagnetic interaction that led to radiative neutron-proton capture in the early universe. The early universe did not need to resort to the comparatively slow weak interaction that initiates the solar cycle. Temperatures in the sun range from about 5,800 K at the surface to about 14 million kelvins at the center, while the temperature of the early universe at 3 min was about 100 million kelvins. This means, as far as the sun is concerned, that once a deuteron has been formed, the ambient radiation is not energetic enough to tear it apart. The average density in the sun is about $1\frac{1}{2}$ times that of water—the central density is about 60 times that of rock, about 156 g cm^{-3}. So, at the center, we are talking about number densities of particles in the range of 10^{29} cm^{-3}, while, as Alpher and Herman showed in the argument we used in the last section, the number density of particles at the time of deuteron formation was only about 10^{17} cm^{-3}. In short, the conditions for element formation by fusion in the sun and other stars are completely different from those of the early universe, so we should not be surprised that the reactions involved are very different as well.

In the sun, the next step after the deuteron is formed involves the electromagnetic reaction $d + p \rightarrow {}^3He + \gamma + 5.49$ MeV, the latter being the energy of the gamma. There is then a branching of possibilities. About 86 percent of the time, the strong reaction ${}^3He + {}^3He \rightarrow {}^4He + p + p$ takes place, leading directly to the helium production we have discussed. But, about 14 percent of the time, the electromagnetic reaction involving one of the available helium nuclei, ${}^3He + {}^4He \rightarrow {}^7Be + \gamma$, can take place. If this happens, there are again two branches to consider. The most likely—about 99.89 percent of the time—is the sequence ${}^7Be + e^- \rightarrow {}^7Li + \nu_e + 0.86$ MeV rapidly followed by ${}^7Li + p \rightarrow {}^4He + {}^4He$. But, and this is the essential chain of reactions needed for Davis's experiment to work, about 0.11 percent of the time there is the reaction ${}^7Be + p \rightarrow {}^8B + \gamma$ followed by ${}^8B \rightarrow {}^8Be + e^+ + \nu_e + 14.06$ MeV and then by ${}^8Be \rightarrow {}^4He + {}^4He$. We have put the maximum neutrino energy in the boron decay because that is crucial for Davis's experiment. Davis's idea was to measure the inverse beta-decay reaction $\nu_e + {}^{37}Cl \rightarrow e^- + {}^{37}Ar$. A bit of study of a table of nuclides will persuade one that the threshold for this reaction is 0.81 MeV, which means that it is insensitive to the neutrinos that come from the initial proton-proton reaction. Moreover, it is just below the 0.86 MeV in the electron capture by beryllium. Hence, *grosso modo*—we have left out some minor additional branches—Davis's detector detects just the beta-decay neutrinos of the boron-8 produced in the sun. To predict how many such events Davis should expect to see in, say, a day requires a very complicated calculation using a solar model. Experts have spent years doing this calculation. One of the most striking things about it is that the neutrino flux from this decay is incredibly sensitive to the central temperature of the sun, T. It goes approximately as T^{18}! The slightest change in the sun's central temperature produces a huge change in the theoretical estimate of the flux. These estimates suggest that Davis should observe, on the average, about one event a day. Rather than giving events per day, the customary unit used here is the so-called solar neutrino unit (SNU), where 1 SNU = one capture per second per 10^{36} target atoms. Translated into SNU, the predicted rate for Davis is given as 6.1 SNU. What about the experimental results? First, how is the experiment done?

The isotope of argon that Davis is looking for—${}^{37}Ar$—is both chemically inert (argon is a noble gas) and radioactive. After being produced, it beta-decays back into ${}^{37}Cl$ with a half-life of 35 days. Once every couple of months, Davis extracts whatever ${}^{37}Ar$ has been produced by neutrinos entering the tank by adding helium to the cleaning fluid, flushing out the argon which is then frozen. The argon decays back into chlorine, and this activity is what Davis measures. Between 1970 and 1990 about 90 runs were made. The number of SNU shows some apparent variation in time which might or might not be significant depending on what the final outcome of Davis's experiment is, but the essential point is that, in magnitude, the SNU number is about one-third of the predicted value. What to make of this and how it might be connected to neutrino masses we come back to shortly, but first we want to describe a complementary experiment that has produced results that appear to be closer to one's theoretical expectations. In the so-called Gran Sasso tunnel, under the Abruzzi Mountains in Italy, an international collaboration of experimental physicists has installed a 50,000-liter vat of gallium chloride. The reason for the tunnel is to provide shielding, and the gallium chloride vat is the neutrino detector. The reaction that is produced this time is $\nu_e + {}^{71}\text{Ga} \rightarrow e^- + {}^{71}\text{Ge}$, where the final nucleus is a radioactive isotope of germanium. This rection has a threshold of only 236 keV, so it is sensitive to all the neutrino-producing reactions in the sun. In June 1992 this experiment, known as the GALLEX collaboration, reported finding 83 SNU. At least one version of the theory predicts a total of 132 SNU-74 coming from the proton-proton reaction, 34 coming from ^{7}Be, and 14 coming from ^{8}B, the rest arising from miscellaneous processes. Hence this experiment is coming closer, although it still seems to be below the theoretical estimates. It is, however, reassuring that GALLEX definitely seems to be observing the basic proton-proton neutrinos. For awhile, when Davis's experiments did not seem to be seeing anything, some people, in desperation, suggested that for some unknown reason the sun might have switched off the weak interactions that produce its energy!

It may be, when all is said and done, that there is no problem and that some refinement of the theoretical model or of the experiment will fix everything. But let us suppose that experiments continue to show a paucity of neutrinos compared to our best theoretical estimates. What might this mean? One of the most interesting suggestions has to do with what are known as neutrino oscillations. This, as we shall see, is where the neutrino masses come into the problem. Particle oscillations—a purely quantum mechanical effect—came to the attention of particle physicists in the late 1950s. They arose in connection with the neutral K-meson system. Let us remind ourselves of what we know about this system from the work of Chap. 5. We said that there was a neutral K meson—the K^0—with a strangeness of $+1$ and a mass energy of 493.646 MeV, along with its antiparticle—the $\bar{K}^0$—which has a strangeness of -1 and the same mass. That was a bit of a fib. In reality, neither of these particles has an exactly well-defined mass. This is not a matter of the uncertainty principle $\Delta E\, \Delta t \sim \hbar$, which tells us that a decaying particle never has a precise mass, but rather of something even more subtle.

Both the K^0 and the $\bar{K}^0$ can decay into a pair of pi mesons. This is a strangeness-violating decay, but weak interactions do violate strangeness. Strong interactions conserve strangeness, and the K^0, for example, is produced typically in a strangeness-conserving strong interaction such as $\pi^- + p \rightarrow K^0 + \Lambda^0$. Thus the neutral K meson that is produced at, say $t = 0$, where t is the time, has a definite strangeness. However, once produced, it makes use of the strangeness-violating decays to transform itself into a $\bar{K}^0$, and then back and forth between the K^0 and the $\bar{K}^0$. However, there are linear combinations of the K^0 and $\bar{K}^0$ wave functions that do not oscillate in this way but rather propagate in time with the usual $e^{-iEt/\hbar}$ factor, where $E = \sqrt{p^2c^2 + m^2c^4}$. It is customary to call these linear combinations K_S^0 and K_L^0. These combinations *do* have definite masses—within the limits of the uncertainty principle—and definite lifetimes. Indeed L and S in the notation above stand for "long" and "short", respectively. The lifetimes of K_S^0 and K_L^0 are very different: $\tau_S = 0.8922 \pm 0.0020 \times 10^{-10}$ s, while $\tau_L = 5.17 \pm 0.04 \times 10^{-8}$ s. At any given time the particle that was created, for example, as a K^0, will then propagate as a linear combination of the form $\alpha e^{-iE_S t/\hbar} + \beta e^{-iE_L t/\hbar}$, where α and β are time-independent complex numbers. If K_S^0 and K_L^0 have different masses, there can be oscillation effects since the probability that, say, a given decay mode will occur depends on the absolute square of this linear combination; i.e., there will be terms of the form $e^{-i(E_S \pm E_L)t/\hbar}$. In fact these oscillation effects have been used to measure the mass difference $m_{K_L} - m_{K_S}$, which turns out to be $3.522 \pm 0.016 \times 10^{-12}$ MeV, an incredibly small number. Hence, when we gave definite masses to K^0 and $\bar{K}^0$, we did not fib by very much.

When Davis's results began appearing and when it began to look as if the neutrinos might actually have small masses, it occurred to people that a version of the K^0, $\bar{K}^0$ scenario might help to explain the experiment. Before we describe this version, which involves oscillations between neutrinos of different families, we need to discuss the experimental limits on the masses of the other two neutrinos, the ν_μ and the ν_τ. The ν_μ mass is measured by measuring the momentum of the mu meson emitted in the decay $\pi^+ \rightarrow \mu^+ + \nu_\mu$. From the two-body kinematics it is easy to see that the unique muon momentum produced by the decay of a pi meson, at rest, is given in terms of the masses by

$$\frac{p_\mu^2}{c^2} = \frac{(m_\pi^2 + m_\mu^2 - m_{\nu_\mu}^2)^2}{4m_\pi^2} - m_\mu^2 \tag{7.3}$$

This is the formula that experimenters use. The problem is that while a priori we do not know the mass of ν_μ, we do expect it to be very much smaller than either m_μ or m_π. Later we shall give a cosmological argument to support this notion. Hence m_{ν_μ} is a very small number that has to be extracted from the difference of, relatively speaking, very large numbers known with a limited precision. To get an accurate measurement of m_{ν_μ} requires a measurement of the momentum to something like 1 part in a 100,000. So far, all the experimenters can be sure of is an upper limit on m_{ν_μ}, which we can write as $m_{\nu_\mu}c^2 < 0.27$ MeV, about half the electron mass. The situation with ν_τ is even worse. Here experimenters study the two-body decay $\tau \rightarrow \pi + \nu_\tau$, this time measuring the momentum of the pion, or rare decays like $\tau^+ \rightarrow K^+ + K^- + \pi^+ + \nu_\tau$. Again, they give an upper limit on m_{ν_τ}, which can be written as $m_{\nu_\tau}c^2 < 35$ MeV. The best limit on the electron neutrino mass is now given as $m_{\nu_e}c^2 < 7.3$ eV. This represents our present knowledge based on laboratory experiments on these masses. Now to family oscillations.

In order to have family oscillations two conditions are necessary:

1. At least two neutrinos of different families must have distinct masses. One of these masses may be zero.
2. There must be a family-changing—presumably weak—interaction that mixes these neutrinos.

For the sake of argument, let us suppose that these conditions are met for the ν_e and the ν_τ. Let us imagine that at $t = 0$ a ν_e is created in a beta decay. It can then begin to convert into a ν_τ, and vice versa. Again we see that we have told a bit of a fib. Neither ν_e nor ν_τ has a definite mass. But there will be linear combinations ν_1 and ν_2 that do. At $t = 0$ we can write the relation between these states and the states corresponding to ν_e and ν_τ in the following way:

$$|\nu_e(0)\rangle = \cos\theta\,|\nu_1(0)\rangle + \sin\theta\,|\nu_2(0)\rangle$$
$$|\nu_\tau(0)\rangle = -\sin\theta\,|\nu_1(0)\rangle + \cos\theta\,|\nu_2(0)\rangle \tag{7.4}$$

The reason for this choice in terms of sines and cosines is the conservation of probability. Particles don't disappear when they oscillate. If we demand that, as states with a definite energy, $|\nu_1(0)\rangle$ and $|\nu_2(0)\rangle$ satisfy

$$\langle\nu_1(0)|\nu_1(0)\rangle = \langle\nu_2(0)|\nu_2(0)\rangle = 1 \tag{7.5}$$

and

$$\langle\nu_1(0)|\nu_2(0)\rangle = 0 \tag{7.6}$$

then both $|\nu_e(0)\rangle$ and $|\nu_\tau(0)\rangle$ will have unit length. Initially, $|\nu_e(0)\rangle$ and $|\nu_\tau(0)\rangle$ will have no overlap, as one can see from Eqs. 7.5 and 7.6. But in the course of time $|\nu_e(0)\rangle$ develops into $|\nu_e(t)\rangle$ given by

$$|\nu_e(t)\rangle = (\cos\theta)e^{-iE_1t/\hbar}|\nu_1(0)\rangle + (\sin\theta)e^{-iE_2t/\hbar}|\nu_2(0)\rangle \tag{7.7}$$

This state does have an overlap with $|\nu_\tau(0)\rangle$, and the absolute square of this overlap is interpreted in quantum mechanics as the probability that a neutrino that was born as a ν_e at $t = 0$ becomes a ν_τ at a later time t. Thus

$$P_{\nu_e\to\nu_\tau}(t) = |\langle\nu_\tau(0)|\nu_e(t)\rangle|^2 = 2\sin\theta^2\cos\theta^2\left[1 - \cos\left(\frac{(E_2 - E_1)t}{\hbar}\right)\right]$$
$$= \sin(2\theta)^2\sin\left(\frac{(E_2 - E_1)t}{2\hbar}\right)^2 \tag{7.8}$$

with $E_i = \sqrt{p^2c^2 + m_i^2c^4}$. We have assumed that at $t = 0$ the neutrinos are born with a well-defined momentum p. This expression has the desirable property that it vanishes when the "mixing angle," theta, vanishes or when $m_1 = m_2$. It also vanishes when $t = 0$. We leave as an exercise for the reader to compute the complementary probability $P_{\nu_e\to\nu_e}(t)$, that is, the probability that a neutrino that was born a ν_e will remain one, and to show that $P_{\nu_e\to\nu_\tau}(t) + P_{\nu_e\to\nu_e}(t) = 1$. That is what we meant when we said that neutrinos can't simply disappear.

Experimentally we are interested in a regime in which $pc/m_i c^2 \gg 1$: the ultrarelativistic regime. This is because the energy release in the decays that produce these neutrinos is much larger than the upper bounds on their rest energies. Thus $E_i \simeq pc + \frac{1}{2}(m_i^2 c^4/pc)$ or

$$P_{\nu_e \to \nu_\tau}(t) \simeq \sin(2\theta)^2 \sin\left[\frac{(m_2^2 - m_1^2)c^4 t}{4\hbar pc}\right]^2 \tag{7.9}$$

Since the neutrino is moving close to the speed of light, the distance it travels in the time t, say r, is given by ct. There are several ways to reexpress Eq. 7.9. Let us call $\Delta m^2 c^4 \doteq |m_2^2 - m_1^2| c^4$. If we measure $\Delta m^2 c^4$ in electronvolts squared, r in meters, and $pc \simeq E_\nu$ in megaelectronvolts, then we can write

$$\frac{\Delta m^2 c^4 r}{4\hbar pc} \simeq 1.27 \Delta m^2 c^4 (\text{eV}^2) \times \frac{r\ (\text{m})}{E_\nu\ (\text{MeV})} \tag{7.10}$$

In terrestrial laboratories there are two classes of experiments we can try to do: "appearance" and "disappearance" experiments. The disappearance experiments are appropriate in situations in which one can make a good estimate of the initial flux of electron neutrinos or antineutrinos produced in, say, a reactor. We see from Eq. 7.9 that at a later time the flux of electron neutrinos or antineutrinos will be reduced by a factor of $1 - P_{\nu_e \to \nu_\tau}(t) \leqslant 1$. This factor depends on where the observer is placed relative to the source. As Eq. 7.9 makes clear, it is an oscillating function of this position. We can imagine situations in which the argument of the sine function (Eq. 7.10) is very large. For example, the apparatus could be very far away from the source and the mass differences not small enough to compensate. In this case, the sine function would undergo an enormous number of oscillations, and the sine squared factor can be replaced by its average, which is one-half. We do not have a good theory to use to predict the mixing angle theta. But if it is on the order of 45°, which would mean complete mixing, then half of the initial flux would disappear. It would oscillate into another neutrino family. In actual reactor experiments, the energy of the antielectron neutrinos produced by the decay of the fission fragments has a maximum of about 8 MeV. This means that these neutrinos, if they do oscillate into another family, do not have the energy to create either a mu or a tau lepton by an inverse reaction. The way an oscillation would be observed in this case is by a reduction of flux. Several experiments of this kind have been done, but none of them have found any evidence for such a flux reduction. This allows one to set limits on the parameters, at least in a limited way. Since we do not know either the masses or the mixing angles a priori, all one can do is to say how large a $\Delta m^2 c^4$ is allowed if a given mixing angle is assumed. Typically, if full mixing is assumed, then $\Delta m^2 c^4 \lesssim 2 \times 10^{-2}\ \text{eV}^2$. High-energy neutrino beams can be manufactured by using the neutrinos from K or pi decay after the K's and pi's have been produced at large energies in an accelerator. These neutrinos do have enough energy to produce the more massive leptons. Experiments with such beams are called appearance experiments because leptons of the "wrong" family appear as a result of putative family oscillations. One can put detectors at various distances from the source in the hope of observing family oscillations. None have so far been observed—although rumors to the contrary surface from time to time—producing limits on the masses that are comparable to the low-energy limits.

From what has been said, the possible relevance of all of this to the neutrinos from the sun should be clear. The electron neutrinos produced in the basic solar energy processes, under the conditions we have specified, might oscillate into other neutrino families, and the flux might therefore be reduced. With the sun there is an additional nuance that has received a great deal of attention in recent years. Neutrinos created deep in the interior of the sun have to travel through a medium of variable density to escape. In effect, the scattering they do produces an index of refraction. It turns out that under the right circumstances there can be resonance phenomena that greatly enhance the oscillations even if Δm^2c^4 is very small. Calculations show that under the right circumstances, even if Δm^2c^4 is as small as $10^{-7}\,\mathrm{eV}^2$, there can be enough oscillation of family by the time a neutrino leaves the sun to explain the missing SNU, if they are missing. We await further word from experimenters.

Once the idea that neutrinos might be massive became respectable again, cosmologists immediately began studying the effect that massive neutrinos might have on cosmology. It has now become a sort of cottage industry. The discussion is usually organized in terms of the following antimonies: "light" versus "heavy" and "stable" versus "unstable". For reasons that we will elaborate on, by "light" we shall mean neutrinos with a mass energy less than 100 eV, and by "heavy" we shall mean neutrinos with a mass energy greater than 1 GeV. In principle, neutrinos with a mass can be unstable. For example, a heavy neutrino ν_H could decay into three light neutrinos or, with a mixture of electromagnetic and weak interactions, into a light neutrino and a photon. The simplest case is that of the light, stable neutrinos: ν_e, ν_μ, and ν_τ. If the neutrinos mix, then the states with a definite mass will be linear combinations of these—ν_1, ν_2, and ν_3. Let us begin with the light-mass case.

From our discussion of helium production we know the history of these neutrinos. Until the universe has cooled down to an equivalent temperature of 1 MeV or so, they are kept in tight equilibrium with the electrons, hence the photons. After that, they decouple and expand freely with the universe. The first question we want to answer is, What does the momentum distribution function of these neutrinos look like after decoupling? To approach this problem, note the following. Suppose we have a function that we can write as $f(Rp)$, where R is the scale factor $R(t)$. Then, clearly,

$$\int d^3pf(Rp) = \frac{1}{R^3}\int d^3pf(p) \tag{7.11}$$

This is just the scaling behavior we want for the number density; i.e., the quantity R^3n is constant as the universe expands. But we know that at decoupling—call the temperature T_d—the neutrino distribution for a massive neutrino is of the form

$$\frac{1}{\exp\left(\dfrac{(p^2c^2+m^2c^4)}{kT_d}\right)^{1/2}+1}$$

This is the equilibrium Fermi distribution. There is no reason to think that anything discontinuous happens at decoupling. Hence we can determine the neutrino distribution after decoupling, $f(Rp)_i$, for the ith neutrino from the combined requirements of scaling and continuity. That is,

$$f\left(\frac{R}{R_d}p\right)_i = \frac{1}{\exp\dfrac{[(p^2c^2(R^2/R_d^2) + m_i^2c^4]^{1/2}}{kT_d} + 1} \tag{7.12}$$

We are assuming here that a neutrino has effectively one degree of spin freedom. This is a tricky point that hinges on the nature of the conventional weak interaction theory. This theory implies that neutrinos are emitted in a predominantly left-handed state even if they have a small mass. Hence such neutrinos have half the number of spin degrees of freedom that, say, electrons do when they interact with photons.

Note that the function given by Eq. 7.12 is of the form $f(Rp)$ and goes over continuously to the equilibrium distribution at $R = R_d$. But it is itself not an equilibrium distribution! The argument that we gave in Chap. 4 (see Eq. 4.51) that the equilibrium distribution is preserved for freely expanding particles does not work when the particles have mass. By the way, it works approximately for nearly nonrelativistic particles if one allows the temperature to vary as $T \sim 1/R^2$.

Why do we care about this distribution? Apart from the fact that it is the distribution the experimenters should find, if they can ever detect the cosmological neutrinos left over from the Big Bang, it also determines the number density of these neutrinos in the present universe. But if the neutrinos have mass, then we can use their number densities to find their contribution to the present energy density of the universe. But, as we shall see, there are limits, given by what we know, as to how much the neutrinos are allowed to contribute to the energy density. They cannot contribute more mass density than is observed. These limits are what will give us what is called the "cosmological" mass limit on the light-mass, stable, neutrinos. The first step, then, in finding this limit is to estimate the number density of a given massive neutrino species in the present universe.

The distribution function given by Eq. 7.12 has the property that it scales with $R(t)$ after decoupling. If we use it to find the number density n_i, which is defined as

$$n_i = \int \frac{d^3p}{(2\pi\hbar c)^3} f\left(\frac{R}{R_d}p\right)_i \tag{7.13}$$

then, as expected, n_i scales as

$$n(T)_i = \frac{R_d^3}{R^3} n(T_d)_i \tag{7.14}$$

We would like to replace the R's here by photon temperatures which are measurable, but we must be careful. As we saw in the discussion following Eq. 6.38 in the last chapter, the quantity that remains constant as the universe expands is RT^{ν} and not RT^{γ}, since when the electrons and positrons annihilate, this heats the photons but not the neutrinos. Decoupling of electrons and neutrinos takes place before this annihilation, so that at decoupling $R_d \sim 1/T_d^\gamma$. But we want to evaluate R in the present, long after annihilation. Hence we must take into account the difference between the electron and neutrino temperatures. Thus, at present, $R \sim 1/T^\nu = (\frac{11}{4})^{1/3}(1/T^\gamma)$ or, from Eq. 7.14,

$$n(T)_i = \tfrac{4}{11}\left(\frac{T^\gamma}{T_d^\gamma}\right)^3 n(T_d^\gamma)_i \tag{7.15}$$

It is at this point that we use the assumption that the neutrinos are "light". One can readily show that

$$\frac{\displaystyle\int_0^\infty dx \frac{x^2}{e^x+1}}{\displaystyle\int_0^\infty dx \frac{x^2}{e^x-1}} = \frac{3}{4} \tag{7.16}$$

Hence the number density of photons at T_d^γ and the number density of effectively massless neutrinos are simply proportional to each other.

The number of degrees of freedom of the photons, counting both polarizations, is two. But this is the same as the number of degrees of freedom of a neutrino plus and antineutrino of a given family, assuming that they each have one effective spin state. With the understanding that the i in $n(T)_i$ takes into account both neutrinos and antineutrinos, we have the following very useful relation

$$n(T)_i = \tfrac{4}{11} \times \tfrac{3}{4}\left(\frac{T^\gamma}{T_d^\gamma}\right)^3 n(T_d)_\gamma \tag{7.17}$$

This relation is even more useful than one might at first think. The factor $(T^\gamma/T_d^\gamma)^3$ scales $n(T_d)_\gamma$, so that we can write Eq. 7.17 in the form

$$n(T)_i = \tfrac{3}{11} n(T)_\gamma \tag{7.18}$$

where, if T refers to the present temperature of the cosmic background radiation, $n(T)_\gamma = 417\ \text{cm}^{-3}$. But notice that after the scaling all reference to the decoupling temperature T_d drops out. This is fortunate because, as we have remarked, different neutrinos decouple at somewhat different temperatures. But this has turned out to be irrelevant in deriving Eq. 7.18. Hence we arrive at the remarkable prediction that each neutrino mass state—counting both the neutrino and the antineutrino—should have at the present time an average number density of $114\ \text{cm}^{-3}$. Thus the total light-neutrino energy density ρ_ν is predicted to be

$$\rho_\nu = \sum_i m_i c^2 \times 114\ \text{cm}^{-3} \tag{7.19}$$

where the sum is over families of neutrinos with a mass m_i. How can we use this result to set limits on the mass sum

$$\sum_i m_i?$$

We can take either what we might refer to as the "radical" approach or the "conservative" approach. In the radical approach we shall assume that the universe is "flat", i.e., that $k = 0$ or, equivalently, that $\Omega = 1$. Moreover, we shall assume that all the missing mass is supplied by the three light neutrinos. There is as yet no experimental data to contradict these assumptions and, if true, it would certainly be an elegant solution to the missing mass problem. This would mean that ρ_ν supplies almost all the critical density $\rho_c = h^2 \times 1.054 \times 10^4 \text{ eV cm}^{-3}$, where $0.4 \leqslant h \leqslant 1$. We saw in the last chapter (in the discussion surrounding Eq. 6.108) that there was an upper bound to the ordinary matter contribution to $\Omega_M = \rho_m/\rho_c \leqslant 0.12$. Let us then suppose that the neutrinos supply the rest. In that case $\Omega_\nu = \rho_\nu/\rho_c \gtrsim 0.88$ or $\rho_\nu \gtrsim 1.5 \times 10^3 \text{ eV cm}^{-3}$ or, from Eq. 7.19,

$$\sum_i m_i c^2 \gtrsim 13.2 \text{ eV}.$$

Thus the radical approach supplies a lower bound on the neutrino masses.

The conservative approach supplies an upper bound. This approach makes use only of things that we think we know about. We think we know that the lifetime of the universe is greater than 10 billion years, and we think we know that the Hubble constant fudge factor is greater than 0.4. If we put these two things together, we can conclude that

$$H_0 t_0 \geqslant 0.42 \tag{7.20}$$

where we have used the fact that $H_0 = 3.5 \times 10^{-18} \text{ s}^{-1} \times h$.

To make use of this, we remind ourselves (see Chap. 2) that for an $\Omega > 1$ universe,

$$H_0 t_0 = \frac{\Omega}{(\Omega - 1)^{3/2}} \cos^{-1}\left(\frac{1}{\Omega^{1/2}}\right) + \frac{1}{1 - \Omega} \tag{7.21}$$

If we substitute the lower bound 0.42 into the left-hand side of Eq. 7.21, we find that this value corresponds to an Ω of 5.88. But an examination of the right-hand side of Eq. 7.21 will convince the reader that it is a monotonically decreasing function of Ω. This means, under the assumptions given, that we have $\Omega \leqslant 5.88$. This is what we shall use to set an upper bound on the neutrino masses.

From the discussion given above we see that we must have

$$\Omega_\nu + \Omega_M \leqslant \Omega \tag{7.22}$$

or

$$\Omega_\nu \leqslant \Omega - \Omega_M \tag{7.23}$$

The arguments of the last chapter gave us an upper bound on Ω_M, but what we need here is a lower bound. Actually, as we mentioned in the last chapter, there is such a bound if one accepts the proposition that one has truly observed cosmological deuterium and lithium. In order for enough of the stuff to have been produced, people who have looked into this question claim that $\Omega_M \geqslant 0.011$. We shall use this number in Eq. 7.23. In other words, we must have $\Omega_\nu \leqslant 5.87$. Thus we have

$$\sum_i m_i c^2 \times 114\,\text{cm}^{-3} \leqslant 5.87 \times 1.054 \times 10^4\,\text{eV cm}^{-3} \times h^2$$

$$\leqslant 6.19 \times 10^4\,\text{eV cm}^{-3} \tag{7.24}$$

Hence

$$\sum_i m_i c^2 \leqslant 543\,\text{eV} \tag{7.25}$$

In the literature one will find smaller bounds than this, but they rest on different assumptions. For example, some authors take a smaller value for the upper bound on Ω. Even our conservative bound is three orders of magnitude better than the terrestrial bound for ν_μ and five orders of magnitude better than the terrestrial bound for ν_τ.

We can also use the neutrino distribution given by Eq. 7.12, $f[(R/R_d)p]_i$, to compute various average quantities. For example, we can compute the average value of the magnitude of the momentum $\langle p \rangle$ of a neutrino of mass m at the present time, assuming these neutrinos have been expanding freely since decoupling. By definition,

$$\langle p \rangle = \frac{\displaystyle\int d^3p \frac{p}{(2\pi\hbar)^3} f\left(\frac{R}{R_d}p\right)_i}{n_i} \tag{7.26a}$$

where we have divided by the number density n_i. By letting $p' = (R/R_d)p$ in the integral, we derive the scaling law

$$\langle p \rangle_i = \frac{R_d}{R} \langle p \rangle_{id} \tag{7.26b}$$

which says that the momentum red-shifts as the universe expands. This was to be expected since the momentum is related to the de Broglie wavelength by $p = h/\lambda$ and the wavelengths are red-shifting. If we recall the connection between the R's and the photon temperatures, being careful to take into account photon reheating, we can write Eq. 7.26 in the form

$$\langle p \rangle_i = \frac{T}{T_d}\left(\frac{4}{11}\right)^{1/3} \langle p \rangle_{id} \tag{7.27}$$

We can make a rough estimate of the average speed at which a light neutrino of mass m_i would move in the present epoch. Anticipating that these neutrinos will be nonrelativistic, we can now write $\langle p \rangle_i \simeq m_i \langle v \rangle_i$. Since the neutrinos are essentially massless when they decouple, we can use the massless approximations to the integrals that determine $\langle p \rangle_{id}$. We leave it as an exercise for the reader to show that $c\langle p \rangle_{id} \simeq 3kT_d$. Thus we have, putting everything together,

$$\frac{\langle v \rangle_i}{c} \simeq 3\left(\frac{4}{11}\right)^{1/3} \frac{kT}{m_i c^2} = \frac{2.1kT}{m_i c^2} \simeq \frac{5 \times 10^{-5}}{m_i c^2/100 \text{ eV}} \tag{7.28}$$

where we have chosen to measure $m_i c^2$ in units of 100 eV. This confirms the fact that any light neutrinos massive enough to close the universe will be at present very nonrelativistic. We have been assuming that after decoupling, our neutrinos freely stream with the expanding universe. It is, however, quite likely that such slowly moving massive neutrinos will become trapped in galactic gravitational fields, hence we might well not find anything like a blackbody distribution for them. In any event, Nobel Prizes are surely to be had for the first experimenters to detect them.

Before we go into the theory of the heavy massive neutrinos, we must first, in this instance, define what we mean by a neutrino. Anything we will call a neutrino must have the same interactions as the light-mass neutrinos we have already studied. If in addition to making the particles massive we also allow them arbitrary interactions, we have a game with no rules. Massive neutrinos can be stable or unstable. We begin by considering the stable case. At first sight, one might think that our previous discussion would have simply ruled out any stable neutrino more massive than about 500 eV. Surprisingly, this is not the case, and this has led to some interesting theoretical work that we only summarize here. By definition of what we call a neutrino, ν_H, the massive neutrino, can interact only with other leptons via the standard weak-interaction cross section which by dimensional analysis has the form $\sigma \simeq G_F^2 E^2/(\hbar c)^4$. In the applications we are making, the energy E will be either kT or mc^2, depending on the circumstances. In the present circumstances, the interaction in question is $\nu_H + \bar{\nu}_H \to l + \bar{l}$, where l stands for any of the light leptons—neutrinos as well as electrons. This is, by assumption, a conventional weak interaction which can proceed, for example, through the exchange of a Z^0. If the mass energy of the ν_H is not larger than $m_{Z^0} c^2 \simeq 91$ GeV, then the local Fermi theory will apply to this annihilation process. For a heavier ν_H, one has to use the theory in which the Z^0 exchange must be taken account of, and the results are different. In the annihilation regime, $E \simeq m_{\nu_H} c^2$, which means that the annihilation cross section increases with increasing m_{ν_H}. This means that the more massive the ν_H are, the easier it is to get rid of them by annihilation. Again at first sight, one might be tempted to conclude that if this is a near-equilibrium process, the number of ν_H would be controlled by the Boltzmann factor $e^{-m_{\nu_H} c^2/kT}$, which would mean that for $kT \lesssim m_{\nu_H} c^2$ the ν_H's would rapidly vanish without a trace. But the annihilation, at least in its later stages, is not an equilibrium process since the cross section is falling as T^2. Furthermore, as the temperature falls below the temperature corresponding to the ν_H rest mass, there will not be enough energy for the inverse processes to take place. Hence there will be a freeze-out of the type we have studied before, and there will be a residual number of ν_H and $\bar{\nu}_H$ pairs remaining. It is these residual pairs that we cannot have too many of in order not to violate the cosmological bound. To sort all this out requires solving the rate equations which in this case one can do with great accuracy.

The solutions show that the freeze-out takes place at about $kT \simeq m_{\nu H}c^2/20$. They also show that the residual energy density $\rho_H(T)$, which includes both neutrinos and antineutrinos, is given approximately in terms of $n(T)_\gamma$, the present photon number density, by the expression

$$\rho_H \simeq n(T)_\gamma \times m_{\nu H}c^2\left(\frac{\text{GeV}}{m_{\nu H}c^2}\right)^3 \times 10^{-9} \tag{7.29}$$

If now we apply the cosmological bound, we find that there is an allowed region for $m_{\nu H}c^2 \gtrsim 2$ GeV. Note that the heavy-neutrino mass appears in the denominator of Eq. 7.29. This somewhat surprising result is why there is a heavy-mass regime in which stable ν_H's do not violate the mass bound. There are just not enough of them around now to cause trouble. Since there is no evidence that these additional heavy families, if they exist, show up in the width of the Z^0, we presumably must have $m_{\nu H}c^2 > 45$ GeV since the Z^0's mass energy is about 90 MeV. The mass window again closes if we allow the mass of the ν_H to get too high. We must then use the theory that takes the Z^0 exchange into account explicitly. When we do this, we can show that there is a new forbidden region when $m_{\nu H}c^2 \gtrsim 10^5$ GeV. But, as we have just seen, there is a very large allowed region of possible high-mass stable neutrinos, at least as far as the mass bounds are concerned.

We must emphasize, however, that no such heavy massive neutrino is known to exist. Nor is there any evidence for the final part of our antimonies—the unstable massive neutrino. Nonetheless, despite the lack of experimental evidence, an enormous amount of theoretical and experimental work has gone into the subject of unstable neutrinos. It would be of fundamental importance to show their existence, if they do exist. There is such an enormous number of constraints and conditions, depending on what sorts of decays are assumed, that we can only scratch the surface of the subject here.

As we mentioned before, one possibility for unstable neutrinos are heavy neutrinos that decay radiatively, namely, $\nu_H \rightarrow \nu_l + \gamma$, where ν_l is one of the light neutrinos. One of the constraints on this mode comes from the cosmic background radiation. The photons produced by neutrino radiative decay could distort the essentially perfect observed blackbody spectrum. This makes it very difficult to accommodate an unstable neutrino that decays in this fashion. More promising, if one is an unstable neutrino enthusiast, are heavy neutrinos that decay into some combination of the three light neutrinos. Here the constraints once again involve the cosmological mass bound. They also involve the lifetime as well as the mass of the neutrinos. These heavy neutrinos can both decay and annihilate. To keep the scenario as simple as possible, one usually assumes that the freeze-out temperature T_f and the temperature at which the particles decay, T_d, are related by $T_d \ll T_f$. This allows a clean separation of the two processes, decay and annihilation. It also allows an interesting new possibility depending on what T_f is. If T_f is less than the 10^4 K at which the universe becomes matter-dominated, the energetic decay products of ν_H will dominate the total energy density. Each of the neutrinos produced in the decay will carry away about $m_{\nu H}c^2/3$ in kinetic energy, which can be very large. But if the decay products are light enough, the universe will once again become radiation-dominated and might still be radiation-dominated. In other words, in this scenario there will be only a brief interval in which the universe is matter-dominated. It is then a good approximation to assume radiation dominance for the entire lifetime of the universe. In that case, we can derive a rather simple approximate formula for the residual energy density which will be dominated by the heavy neutrino's decay products, namely,

$$\rho_{\mathrm{H}} \simeq \pi^{1/2} m_{\nu_{\mathrm{H}}} c^2 \times n(T^{\gamma})_{\gamma} \left(\frac{T^{\gamma}}{T_f}\right)^3 \left(\frac{\tau_{\nu_{\mathrm{H}}}}{t_0}\right)^{1/2} \tag{7.30}$$

Here T^{γ} and t_0 stand for the present photon temperature and lifetime, respectively, while T_f stands for the freezing temperature and τ_{ν_H} stands for the heavy-neutrino lifetime. If we fix a mass for the neutrino, we can use Eq. 7.30 to set an upper bound on the ratio of the lifetime of the neutrino to that of the universe. For example, a 1-GeV neutrino has a maximum lifetime of about $10^{-4} t_0$. The reader can try other combinations. Again, we have no evidence that such neutrinos exist.

The last topic we want to take up in this chapter is not precisely cosmology, although it certainly is cosmic. At 7:35:41 Greenwich mean time, on February 23, 1987, during an interval of 5.59 s, eight neutrinos—actually, it turned out, antineutrinos—were registered in a detector located 2,000 feet down a salt mine about 20 miles from Cleveland, Ohio. The detector was a tank of superpurified water. The tank, which is known as the IMB (University of California at Irvine, University of Michigan, and Brookhaven National Laboratory) has a 60-foot width, a 65-foot depth, and an 80-foot length. It holds about 8,000 metric tons of water. The water is so clear that when divers first went down into the tank to check the photomultiplier tubes that line it, they got vertigo. The tank was completed in 1980, and its purpose was to detect the decay products of the proton, if any. Up until recent years, the proton was taken to be an absolutely stable particle—something that was canonized into the law of conservation of baryons. Since there is no baryon lighter than the proton, baryon conservation guaranteed its absolute stability. But with the success of the unified theories of weak and electromagnetic interactions, it seemed like an attractive idea to create superunified theories that would unite baryons and leptons. One consequence of these theories is the possible instability of the proton. It can have decay modes such as $p \rightarrow e^+ + \pi^0$, and it was decays of this sort that the tank was designed to detect. Up to the present, none have been found, and the proton lifetime is thought to be greater than about 10^{32} yr. But the tank is also a very good detector of antineutrinos. It contains over 10^{33} protons, and these can undergo the inverse beta-decay reaction $\bar{\nu}_e + p \rightarrow e^+ + n$. It is the positron that the phototubes detect, after it has annihilated with an electron. What the IMB detector had detected were the electron antineutrinos produced in the explosion of a supernova located in the relatively nearby Large Magellanic Cloud, which is estimated to be about 55 kpc away. The supernova, which came to be known as SN1987A, was so close that, for the first time since a similar event occurred during the lifetime of Kepler, one could see it with the naked eye. At the same time, 11 additional electron antineutrinos arrived in a period of 12.439 s at a similar detector, this one in a zinc mine located in Kamioka, Japan, called Kamiokanda II. Never in the history of astrophysics have so many astrophysicists owed so much to so few events.

Before its explosion, SN1987A was an unprepossessing supergiant blue star—Sanduleak-69202—so named because it was in the star catalog of Nicholas Sanduleak of Case Western Reserve University. Like other such stars, it had a dense core of iron—the heaviest element that can be produced in stellar fusion processes—surrounded by onion skin layers of silicon, oxygen, carbon, helium, and hydrogen. When the core became too massive, it collapsed in a few milliseconds. It released about 3×10^{53} ergs of energy in about a second. Our peaceful sun releases energy at a rate of "only" about 3.9×10^{33} ergs sec^{-1}. Most of the supernova explosion energy is released as neutrinos and antineutrinos—about 10^{58} in all. Of these, about 30 million billion of them passed through the IMB detector in roughly 6 s, producing eight detectable interactions. It immediately occurred to cosmologists and others that as few as these events were, they could be a goldmine of information about the $\bar{\nu}_e$, which can produce positrons in an inverse beta-decay reaction.

Consider, for example, its mass. To see how this is relevant, we will use the fact that the neutrinos that arrived from the supernova had a spectrum of energies. The IMB antineutrinos ranged in energy from 20 to 40 MeV, while the Kamiokanda antineutrinos ranged from 7 to 50 MeV. In general, the mass of a neutrino—or antineutrino—and its speed are related. Indeed,

$$\frac{v}{c} = \frac{pc}{E} = \sqrt{1 - \frac{m^2c^4}{E^2}} \simeq 1 - \frac{1}{2}\frac{m^2c^4}{E^2} \tag{7.31}$$

The latter approximation takes advantage of the fact that the mass energy of the $\bar{\nu}_e$ is much less than its kinetic energy. The fact that the neutrino has a mass is what produces a spread in the time of arrival of the neutrinos of different energies, assuming they were all emitted at the same time to begin with. We leave it as an exercise for the reader to show that if two neutrinos of mass m have energies E_1 and E_2 and travel a distance d, then their mass energy m is related to the difference in their time of arrival, Δt, assuming they start at the same time, by the equation

$$m^2c^4 \simeq \frac{2c\,\Delta t}{d} \times \frac{E_1^2E_2^2}{E_1^2 - E_2^2} \tag{7.32}$$

If we put the Kamiokanda parameters into this equation, we find that $m_{\nu_e}c^2 \simeq 21$ eV. This is presumably an upper bound to the mass since part of the Δt could reflect a time spread in emission of the neutrinos. The fact that the neutrinos arrived at all has also been used to set limits on the lifetime of the $\bar{\nu}_e$. This seems like a somewhat academic exercise since most particle physicists are persuaded that if the neutrinos do have a mass, the electron neutrino will be the lightest and so it is not clear what it could decay into. But if we take a purely empirical point of view, we can set limits on the lifetime by noting that it takes a time

$$\frac{d}{v} \times \frac{1}{(1 - v^2/c^2)^{1/2}}$$

to get here, taking the time dilatation into account. We again leave it as an exercise for the reader to show that this sets a lower limit on the ratio of the lifetime to the mass. In brief, the supernova has been a wonderful laboratory. We hope we won't have to wait another 500 yr for the next visible explosion.

Up until now in this book Einstein's theory of general relativity has been lurking in the background like some sort of majestic ghost. In the next chapter we shall discuss some of its implications for the propagation of light.

8

Horizons

As we mentioned at the end of the last chapter, it is remarkable how far one can get with what we might call "physical cosmology" without introducing the formalism of general relativity. This is basically due to the simplifying assumptions we have made about the gross structure of the universe. The two most important of these assumptions are homogeneity and isotropy. As we have already emphasized, these assumptions are supposed to hold only in an average sense since it is obvious when we look around us that, at least in our immediate neighborhood, the universe is neither homogeneous nor isotropic. The moon, for example, is at any given time on one side of the earth or the other. On the other hand, the cosmic background radiation appears to be nearly homogeneous and, with some exceptions that we think we understand and will later discuss, isotropic. With these assumptions, the universe is about as Newtonian as general relativity allows. However, the one place where we cannot ignore the effects of general relativity is in the propagation of light. There is nothing more relativistic than light propagation. It is here that we must bring in general relativity.

As we pointed out in the discussion surrounding Eq. 1.30, the key quantity in general relativity is the metric tensor $g_{\mu\nu}(x, y, z, t)$. This function of space and time determines, for example, how light propagates in curved space. Given $g_{\mu\nu}$ we can form, as we did in Eq. 1.30, the metric ds^2:

$$ds^2 = \sum_{\mu,\nu=1}^{4} g_{\mu\nu}\, dx^{\mu}\, dx^{\nu} \tag{8.1}$$

In general, $g_{\mu\nu}$ will be a complicated function of space and time with off-diagonal elements that make the metric separation of space and time impossible. However, with our assumptions, as we will now show, the metric becomes enormously simplified. To apply homogeneity and isotropy to the problem of simplifying the metric we must first be a little more precise by what we mean by these terms. We certainly do not mean that the universe looks the same in every direction to every observer at all times. We know, to take an obvious example, that the photon density decreases with time. To see what we do mean, we can proceed in the following somewhat fanciful way.

We may imagine that we are on the surface of the earth and that it begins to expand. We might find it useful, as the expansion evolves, not to change our street addresses. These coordinate designations—our street addresses—would not then be a function of time. They would be what is called comoving with respect to the expansion. This would not mean that the distances between houses would not increase. They would, but the house coordinates would retain, by convention, the same values at all times. We may imagine that each house is supplied with an identical clock and that, by means of light signals, these have been synchronized. The "proper time" of all the houses would be the same as time went on. It would then be meaningful to ask each household to keep a diary of observations made in all directions at any given proper time. If these diaries, upon examination, turned out to be identical, our earthly universe would by definition be homogeneous and isotropic. We can now imagine generalizing this procedure to points all over the universe. Once again, comparing diaries, if they are all the same, then our universe is by definition homogeneous and isotropic. The simplicity of the metric for this universe manifests itself when these comoving coordinates are chosen. This is what we now want to demonstrate.

The first thing we wish to show is that for this simplified universe there are no off-diagonal terms in $g_{\mu\nu}$. We shall give the argument that proves that there are no terms of the form, say, $dx\,dt$, and we shall leave as an exercise the proof that there are terms of the form, say, $dx\,dy$. To this end, we need a fact from the general theory of relativity. To appreciate this fact, let us return to flat space. Let us also set $dx = dy = 0$ so that $ds^2 = dx^2 - c^2\,dt^2$. If we now set $ds^2 = 0$, we have $dx/dt = \pm c$. But this is the equation of a light ray propagating along the positive or negative x axis. It is this result that generalizes. In general relativity, light propagates along a path that satisfies $ds^2 = 0$. Such a path is called a "null geodesic"—a geodesic being the shortest distance between two points in space-time where distance is defined in terms of the metric tensor $g_{\mu\nu}$. Let us suppose we have a metric of the form $ds^2 = A\,dx^2 + B\,dx\,dt - Cc^2\,dt^2$, where A, B, and C are arbitrary functions of space and time. If we set $ds^2 = 0$ for this metric, there will be in general two very different solutions for dx/dt. This differs from the flat space case where the two solutions simply represent the fact that light can travel from A to B or from B to A along the x axis, in a perfectly symmetric manner, with the speed c. In this new case with the $dx\,dt$ term, the symmetry is destroyed. We can connect A to B with one solution, and B to A with another. The two observers will report differing light propagation—even the light speeds will be different—and the homogeneity will be destroyed. Hence, assuming the result of the exercise that shows that terms like $dx\,dy$ are also not allowed, we have reduced the metric to the form $ds^2 = A\,dx^2 + B\,dg^2 + C\,dz^2 - D\,c\,dt^2$.

To make the next simplification, we recall that in a frame of reference at rest with respect to an observer—one of the houses in our comoving system—$dx = dy = dz = 0$. For this observer $ds^2 = -c^2\,d\tau^2$, where τ is the proper time. Hence we have reduced the metric to the form $ds^2 = A\,dx^2 + B\,dy^2 + C\,dz^2 - c^2\,dt^2$, which reduces to $ds^2 = -c^2\,d\tau^2$ in the rest frame. The choice of the minus sign is conventional. To make the next step, we recall that to preserve the isotropy as the universe expands—to allow triangles to evolve into similar triangles—we had to demand that the dimensions scaled with a common factor $R(t)$. This means that we can write the metric in the form $ds^2 = R(t)^2\, dl^2 - c^2\, dt^2$, where dl^2 is a function of the spatial variables alone. The hard part of the job is determining the form of dl^2.

A reader of this book can find several others dealing with the formalism of general relativity where this dl^2 is derived with varying degrees of rigor. The result of these derivations, all of which use some of the machinery of the general theory of relativity, goes under the rubric "Robertson–Walker metric" after the first people, H. G. Robertson and A. G. Walker, to write it down in the late 1920s using these general arguments. The metric had been used in Friedmann's earlier papers, but he did not discuss its generality. We will spare the reader these formal arguments and approach the task by means of a charming expanding universe model invented by the British cosmologist E. A. Milne in 1932. It is referred to as the Milne "dust model".

In the Milne dust model, one supposes that there was a creation act (we don't inquire why) at a time $t = 0$ in which dust was spewed out of a source at $x = y = z = 0$ in some fundamental coordinate system which we will call S. This dust, it is assumed, is ejected with all possible speeds less than c, and in all possible directions. We also assume that the dust particles do not interact with each other gravitationally, nor do they collide. They simply expand freely, retaining their initial velocities. If the distance to an arbitrary dust particle is, say $\mathbf{r}$, then we clearly have $\mathbf{r} = \mathbf{v}t$, where $\mathbf{v}$ is the speed at which the particle was ejected at $t = 0$. But this is just the Hubble law from the point of view of an observer at the origin of S. That is, we can solve for $|\mathbf{v}|$ in terms of $|\mathbf{r}|$ and find

$$|\mathbf{v}| = \frac{|\mathbf{r}|}{t} \tag{8.2}$$

which is Hubble's law with $R(t) = At$, where A is a constant. Now let us consider an arbitrary dust point P located at $\mathbf{r}_P$, along with a second dust point located at $\mathbf{r}$; see Fig. 1. We have also drawn a vector $\mathbf{r}_q$ from P to the second dust point.

Associated with these dust points are three velocities that we can call $\mathbf{v}$, $\mathbf{v}_P$, and $\mathbf{v}_q$. We see that

$$\mathbf{r}_q = \mathbf{r} - \mathbf{r}_P = (\mathbf{v} - \mathbf{v}_P)t = \mathbf{v}_q t \tag{8.3}$$

What this equation says is that, viewed in S, an observer at P will also report a Hubble law with the expansion apparently centered around P! If we transform to

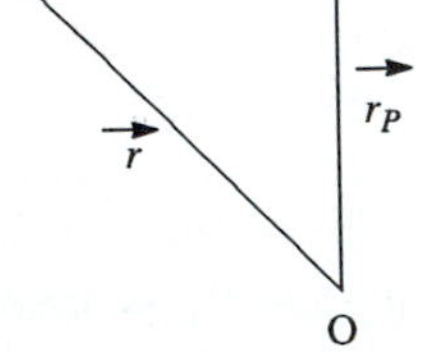

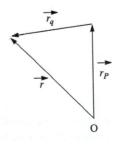

Figure 1 Three dust points as located by an observer in S.

a system comoving with P, then this comoving observer will report a Hubble law with the expansion centered around himself. Hence the Milne dust model is a model of an expanding universe obeying a Hubble law. But we want to make sure that we can doctor the model so that it is homogeneous and isotropic.

To this end, we go to the system comoving with P and invite the observer to look around at any proper time τ. There isn't anything to observe except dust, which is characterized by a density at P which we call $n_P(\tau)$. As time passes, this density increases according to $n_P(\tau) = B/\tau^3$, where B is a constant. The factor of τ^{-3} arises since all the linear dimensions are increasing as τ. To ensure homogeneity we demand that, in our universe, each observer assigns the same constant B to the density. At first blush, we might think that this model universe has a boundary composed of those dust particles that actually move with the speed of light. In what sense this is a boundary is somewhat a matter of interpretation. Certainly there can in principle be observers outside the advancing dust cloud. These observers will see a dust cloud with a boundary and can, if they are so inclined, interact with the dust particles within. On the other hand, as we will now argue, viewed from the inside, the universe does not really have an "edge". To see what is meant, let us study the density $n_p(\tau)$ from the point of view of the observer at the origin in S. We must make a Lorentz transformation in which the translation velocity is $|\mathbf{r}_P|/t$. The connection between the proper time τ and the time t of the observer at the origin in S is given by $t = \tau/\sqrt{1 - |\mathbf{r}_P|^2/c^2t^2}$. We will use this to transform the τ^{-3} factor in $n_P(\tau)$. But there is another consideration in making this transformation. A density goes as a total number N, which is a Lorentz invariant—all observers will agree on the total number—however, divided by a volume V, which isn't; that is, $n \sim N/V$. But, under a Lorentz transformation, we have $V' = V\sqrt{1 - v^2/c^2}$. Only one factor of the square root enters into this transformation of the volume V since we can imagine transforming along, say, the x axis, which will leave the y and z components unaffected. Thus, viewed from the origin, $n_P(\tau)$ takes the form

$$n_P(t)' = \frac{At}{(t^2 - r^2/c^2)^2}. \tag{8.4}$$

But at the edge, $r = ct$, and, as viewed from S, the density at P becomes infinite. This presumably means that the edge is unattainable as viewed from S. But this same argument can be made if we view the density at $\mathbf{r}$ from P. This density will also blow up at the edge, so we can say, if we like, that there is no edge since all the observers will agree that it is unattainable. This is one of the special features of the Milne dust model. Next, we turn to the matter of metrics.

Since there is no gravitation in the Milne dust model, we are perfectly welcome to use the flat space metric in S:

$$ds^2 = dx^2 + dy^2 + dz^2 - c^2\,dt^2 = dh^2 + h^2(d\theta^2 + \sin\theta^2\,d\phi^2) - c^2\,dt^2 \tag{8.5}$$

We have used h for the radial coordinate, saving r for the eventual comoving radial coordinate. Since the dust particles are in motion, these coordinates change in time.

To put the metric into its "canonical" form, $ds^2 = R(t)^2\,dl^2 - c^2\,dt^2$, we must transform to comoving coordinates. The following transformation will do the trick:

$$h = cr \sinh \chi \qquad t = \tau \cosh \chi \tag{8.6}$$

where χ is any number between zero and infinity. We can now make use of the relations

$$\frac{d}{d\chi}\sinh \chi = \cosh \chi \tag{8.7a}$$

$$\sinh \chi^2 = \cosh \chi^2 - 1 \tag{8.7b}$$

to rewrite Eq. 8.5 in the form

$$ds^2 = c^2\tau^2[d\chi^2 + \sinh \chi^2(d\theta^2 + \sin \theta^2\, d\phi^2)] - c^2\,d\tau^2 \tag{8.8}$$

In this way of writing the metric, χ, θ, and ϕ are the fixed comoving coordinates and the scale factor $R(\tau)$ is given by $R(\tau) = c\tau$. We can also rewrite the metric by introducing a "radial" coordinate r by the definition

$$r = \sinh \chi \tag{8.9}$$

Since

$$\frac{d}{d\chi}\sinh^{-1} r = \frac{1}{(1+r^2)^{1/2}} \tag{8.10}$$

we have

$$ds^2 = c^2\tau^2\left[\frac{dr^2}{1+r^2} + r^2(d\theta^2 + \sin\theta^2\, d\phi^2)\right] - c^2\,d\tau^2 \tag{8.11}$$

Here again, r is a coordinate whose value runs between zero and infinity. It is the form of this metric that we shall use shortly.

Since our principal motivation for introducing the general theory of relativity was to discuss the propagation of light, this is where we shall begin. In Chap. 2 of the Macropaedia, we gave an argument (see the discussion following Eq. 2.6) that purported to show that light in an expanding universe is red-shifted. A discriminating reader might have felt some dissatisfaction with this account since it mixed relativistic and nonrelativistic notions. We now propose to remedy this. The argument we are going to give dates back at least as far as Lemaître's 1931 paper. We begin by presenting an approximate identity of the integral calculus whose proof we leave as an exercise for the reader:

$$\int_{x_1+\delta x_1}^{x_2+\delta x_2} f(x)\,dx - \int_{x_1}^{x_2} f(x) = \delta x_2 f(x_2) - \delta x_1 f(x_1) + O(\delta x_2^2, \delta x_1^2) \tag{8.12}$$

We shall apply this identity in the following way. As we have mentioned, in general relativity light travels along a null geodesic—a "straight line"—determined by the condition $ds = 0$. To simplify things without changing anything essential let us suppose that we are interested in light that travels along a radial direction from one observer to another. We endow our "dust" particles with the property of emitting and absorbing light. Thus, in the Milne model, light propagating in the radial direction has (see Eq. 8.11) its trajectory determined by the equation

$$\frac{d\tau}{\tau} = \frac{dr}{(1+r^2)^{1/2}} \tag{8.13}$$

Both the integrals on the left- and right-hand sides of this equation can and will be done, but we now want to make a general argument in which Eq. 8.13 will be a special case. Let us rewrite it in the form

$$c\frac{d\tau}{R(\tau)} = dr\,f(r) \tag{8.14}$$

If we call the time of emission of light from a dust particle with its radial coordinate r_1, τ_1, and if we call the time of absorption at r_2, τ_2, we have

$$c\int_{\tau_1}^{\tau_2} \frac{d\tau}{R(\tau)} = \int_{r_1}^{r_2} f(r) \tag{8.15}$$

But we can imagine that the emitter emits a second pulse at $\tau_1 + \delta\tau_1$ which is absorbed at time $\tau_2 + \delta\tau_2$. Since the coordinates are comoving, the r integral is the same in both cases. Hence the left-hand side of Eq. 8.12 vanishes, and we have

$$\frac{\delta\tau_1}{R(\tau_1)} = \frac{\delta\tau_2}{R(\tau_2)} \tag{8.16}$$

or

$$\frac{\delta\tau_1}{\delta\tau_2} = \frac{R(\tau_1)}{R(\tau_2)} \tag{8.17}$$

If we think of $\delta\tau$ as the time interval between pulses of light—inverse to the frequency—then $c\delta\tau = \lambda$, the light's wavelength. Hence under the condition that the wavelength of the light is much shorter than the distance the light has traveled—$\lambda \ll c\tau$—we have the red-shift formula

$$\frac{\lambda_1}{\lambda_2} = \frac{R(\tau_1)}{R(\tau_2)} \tag{8.18}$$

This is a more satisfactory derivation than the previous one, and makes the underlying assumptions much clearer. One may worry, however, that this derivation is an artifact of the change in variables we made to arrive at the metric given by Eq. 8.11. In general relativity, it is often possible to induce remarkable and specious effects by redefining the metric. As we will now show, this is not the case here. We can rewrite Eq. 8.18 as

$$\frac{\lambda_2 - \lambda_1}{\lambda_1} = \frac{R(\tau_2)}{R(\tau_1)} - 1 \tag{8.19}$$

In the Milne model the right-hand side of this equation is simply $\tau_2/\tau_1 - 1$. If we work to order v/c, $\tau \simeq t$. Hence the right-hand side can be written as $(t_2 - t_1)/t_1$. But the numerator is just the distance traveled by the light in the time interval $t_2 - t_1$. This divided by t_1 is just the Hubble speed v. Hence, in terms of the original time-dependent flat space coordinates, we have a first-order Doppler shift given by v/c, where v is the Hubble speed. We have made the coordinate transformation to the comoving coordinates because we wanted to put the metric into a canonical form that generalizes to the case in which gravitation is included. When gravitation is included, one cannot transform to flat space "globally". Gravitation really curves space, or better, space-time. The best one can do is to make local transformations to flat space in a small neighborhood of a given space-time point.

One of the things we can use the comoving metric for is to consider the matter of horizons. A horizon represents the limit of our vision. Some events that have happened in the universe may never be visible to us even if we could live forever. They are beyond what is known as the "event horizon". This horizon separates those photons that take a finite time to reach us and those that take an infinite time. The event horizon consists of those photons that reach us at $\tau = \infty$, providing the universe still exists at $\tau = \infty$. To determine the radial coordinate of a point on the event horizon we may integrate both sides of Eq. 8.14 and take the limit and τ goes to infinity. Thus, calling a point on the event horizon r_h and the arbitrary initial space and time points r_i and τ_i, respectively, we have by the definition of r_h

$$\lim_{\tau \to \infty} \int_{\tau_i}^{\tau} d\tau' \frac{c}{R(\tau')} = \int_{r_i}^{r_h} dr' f(r') \tag{8.20}$$

The limit on the left-hand side of Eq. 8.20 may be finite or infinite. If it is infinite, we must have r_h infinite since that is the only way, barring pathologies, that the right-hand side can also be infinite. If r_h is infinite, the event horizon is at infinity and eventually we will see everything.

In the Milne model the integrals are easy to work out. We have

$$\lim_{\tau \to \infty} \ln\left(\frac{\tau}{\tau_i}\right) = \ln\left[\frac{(r_h^2 + 1)^{1/2} + r_h}{(r_0^2 + 1)^{1/2} + r_0}\right] \tag{8.21}$$

The right-hand side can also be written in terms of $\sinh^{-1} x$, but this form may be little more transparent. Clearly, the left-hand side goes to infinity, so in the Milne model the event horizon is infinite. It is also clear that we can in principle find other model universes in which the event horizon is finite.

Cosmologists also discuss something they call the particle horizon. These are points that can be reached by photons that started off at the "creation" and have traveled until the present time τ_0. If the universe were flat and not expanding, these points, r_p, would lie on a surface given by $c\tau_0$. In the more general case r_p is determined by the integral

$$\int_{\tau_i}^{\tau_0} d\tau' \frac{c}{R(\tau')} = \int_0^{r_p} dr' f(r') \tag{8.22}$$

where we have taken, by convention, the initial coordinate to be located at the origin. If the left-hand integral blows up at its lower limit, which could happen if we take the lower limit to be zero, then there is no particle horizon; otherwise, there is one. In the Milne model, according to this definition, there is no particle horizon. However, the Milne model is really a flat space model and, written in terms of the original coordinates, a more sensible statement would seem to be that the particle horizon is at the boundary of the dust cloud.

Up to this point in the book, when discussing distances, we have acted as if space is flat and the geometry Euclidean. We are now in a position to remedy this. The first problem we confront is what is meant by the term "distance" in cosmology. The simplest and perhaps most fundamental definition is also the least practical. It is what is called the "proper distance". We introduce the idea in the context of the Milne model, and it will be easy to generalize it. We recall that, written in terms of the comoving coordinates, the metric in the Milne model takes the form $ds^2 = c^2\tau^2[dr^2/(1+r^2) + r^2(d\theta^2 + \sin\theta^2\, d\phi^2)] - c^2\, d\tau^2$. We shall define the "radial" proper distance by making a measurement in a fixed direction with $d\theta = d\phi = 0$. Moreover, we shall do this at a fixed time τ so that during the measurement $d\tau = 0$. How one could possibly organize such a cosmic conspiracy is very unclear. That is why this is not a very practical definition. In any event, the proper distance d_P between two radial comoving coordinates, say 0 and r, is given in the Milne model by integrating ds while setting $d\tau = d\theta = d\phi = 0$. We find at once that

$$d(r)_P = c\tau \sinh^{-1} r \simeq c\tau r + O(r^3) \tag{8.23}$$

where we have used the alternate form of the radial integral. One thing to note about this expression is that it clearly reflects the fact that, although the comoving coordinates never change their values in the course of time, the "distance" between two such coordinates is truly expanding, in this case linearly with τ.

In Chap. 1 of the Macropaedia we presented a discussion of how one really goes about the business of actually measuring distances to cosmological objects. However, this entire discussion was transacted as if space were Euclidean and nonexpanding. We can now remedy this. We shall do so in the context of the Milne model and later generalize. As a preliminary, note the following. Suppose we take a point $r = 0$ and draw a spherical surface around it, at some proper time τ, that intersects a set of points that have a common radial coordinate r. What is the surface area of this sphere? Referring once again to the Milne metric, we see that it is given by $A = 4\pi r^2 c^2 \tau^2$. Not to be confused by the dimensions, remember that r is dimensionless. Now let us imagine that we have a photon emitter at $r = 0$ that every $\delta\tau_e$ seconds emits a photon of frequency ν_e in a random direction. Then the absolute luminosity—the emitted energy per second—of this photon emitter, L_A, is given by $h\nu/\delta\tau_e$. In Chap. 1 (Eq. 1.14) we introduced the definition of the areal "power" P—the energy per second per unit area—that an observer at distance d from this source would measure. The relation was $P = 1/4\pi d^2 \times L_A$, from which a so-called luminosity distance d_L could be defined, namely, $d_L = \sqrt{L_A/4\pi P}$. This is the relation that is used to measure, for example, the distance to the Cepheid variables. We now want to see how it is modified in our curved and expanding space. There are three effects. Let us suppose that we receive one photon every $\delta\tau_0$ seconds at our detector located on the spherical shell of radius $rc\tau$. In a flat stationary space we would have, if the emitting and absorbing oscillators were identical, $\delta t_0 = \delta t_e$. But using Eq. 8.17, we have in the Milne space, $\delta t_0 = \delta t_e \tau_0/\tau_e$. That is the first effect. The second effect is the red-shifting of the energies, namely, $\nu_0 = \nu_e \tau_e/\tau_0$. Finally, we must use the modified surface area of the sphere whose surface passes through points with a comoving radial coordinate r. Putting these things together we have, in the Milne model, the relation

$$P = \frac{1}{4\pi} \times \frac{1}{r^2} \times \frac{1}{c^2\tau_0^2} \times \frac{\tau_e^2}{\tau_0^2} \times L$$

where L is the absolute luminosity (the energy emitted per second) measured by an observer at the emitter. If we use the same definition of the luminosity distance as before, we find that $d_L = rc\tau_0 \times \tau_0/\tau_e$. Referring to Eq. 8.23, we see that the luminosity and the proper distances approach each other as r becomes small and $\tau_0/\tau_e \to 1$. To see what is big and what is small, we must reexpress r in terms of observables. To this end, we use the relation (see Eq. 8.21)

$$\frac{\tau_0}{\tau_e} = (r^2 + 1)^{1/2} + r \tag{8.24}$$

We can invert this to find

$$r = \left(\frac{\tau_0^2}{\tau_e^2} - 1\right)\frac{\tau_e}{2\tau_0} \simeq \frac{\tau_0 - \tau_e}{\tau_e} \tag{8.25}$$

where the approximate equality holds when $\tau_0/\tau_e \sim 1$. We can reexpress the luminosity distance using Eq. 8.25. Thus

$$d_L = \frac{c\tau_0}{2}\left(\frac{\tau_0^2}{\tau_e^2} - 1\right) \simeq c\frac{(\tau_0 - \tau_e)\tau_0}{\tau_e} \simeq d_P \tag{8.26}$$

The final approximate relationship follows from Eq. 8.23.

We could now go through each of the distance measurement techniques discussed in Chap. 1 of the Macropaedia to see how they are modified in the presence of a curved and expanding space, but we shall not do that here. Rather we shall look at another possible definition of distance, the so-called angle distance d_A, which we will need in the next chapter. Let us suppose we want to know the distance to an astronomical object—a galaxy say—that subtends a measurable small angle δ when we observe it. We suppose all the points of this structure are at comoving coordinate r and proper time r; see Fig. 2. We see from the Milne metric that the proper arc length l_A is related to δ and r by the expression $l_A = c\tau r\delta$. What is called the angle distance d_A is given by

$$d_A = \frac{l_A}{\delta} = c\tau r \tag{8.27}$$

We see that once again, for small r, then angle distance approaches the proper distance d_p. In fact, if we worked it through, we would find that all the distance measurements we defined earlier in the book approach the proper distance for small

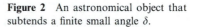

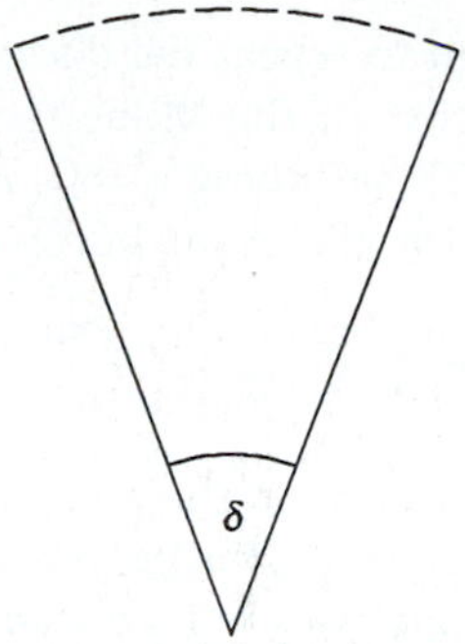

Figure 2 An astronomical object that subtends a finite small angle δ.

r. For large r there will be significant corrections, and these will depend on the cosmological model employed.

As charming and transparent as the Milne model is, it does not represent the real world. To take the next step into the real world we must do something that we said we would do very little of in this book, namely, resort to "It can be shown that". To show these things in detail would require a longish detour into the formalism of general relativity. But the Milne model will make them plausible. The first thing we wish to write down, invoking "It can be shown", is the most general form of the Robertson–Walker metric, written in terms of comoving coordinates r, θ, and ϕ:

$$ds^2 = R(\tau)^2\left[\frac{dr^2}{(1-kr^2)} + r^2(d\theta^2 + \sin\theta^2\, d\phi^2)\right] - c^2\, d\tau^2 \tag{8.28}$$

where k can have the values ± 1 or 0 and $R(\tau)$ is the arbitrary scaling factor. Thus the Milne model is a special case of the Robertson–Walker metric with $k = -1$ and $R(\tau) = c\tau$. In fact every $k = -1$ matter-dominated model approaches the Milne model asymptotically in time. This is plausible since, as $R(\tau)$ approaches infinity, the density of gravitating matter approaches zero. The second thing we want to claim—and this follows from Einstein's general relativity equations (the ones that replace Newton's law $F = ma$)—is that the k that occurs here is precisely the same k that occurred in the Friedmann equation

$$\dot{R}(\tau)^2 + kc^2 = \frac{8\pi}{3} \times G\rho(\tau)R(\tau)^2 \tag{8.29}$$

which we derived from a Newtonian point of view in Chap. 2 of the Macropaedia. This equation, it can be shown, also follows from Einstein's theory when we use the metric given by Eq. 8.28. The Milne model then represents a special case in which the gravitational constant G has been set equal to zero so that $\dot{R}(\tau) = c$. The dust particles do not interact gravitationally. We see, by the way, that in the Milne model space expands at the speed of light. We might worry that this violates relativity. However, as Einstein was fond of saying, "Space is not a thing!" There are no restrictions on the magnitude of $\dot{R}$, and when we discuss what is known as "inflation", we will see that there may well have been an epoch when the universe expanded *faster* than the speed of light.

We are now in a position to repeat the discussion of the horizons and distances that we gave in the special case of the Milne model, in general. We begin with the horizons. We modify Eq. 8.21 by taking $r_i = 0$, which we can always do. This leads to the following expression for the event horizon r_h,

$$\lim_{\tau \to \infty} \int_{\tau_i}^{\tau} d\tau' \frac{c}{R(\tau')} = \frac{\sin^{-1}\sqrt{k} \times r_h}{\sqrt{k}} \tag{8.30}$$

This compact expression includes all three cases $k = \pm 1$ and 0. To carry out the integral on the left-hand side we need a theory of the expansion. But we can recast the integral in a form that allows us to make some general statements. We may write it as

$$\int_{\tau_i}^{\tau} c \frac{d\tau'}{R(\tau')} = c \int_{R(\tau_i)}^{R(\tau)} \frac{dR}{R\dot{R}} \tag{8.31}$$

We can now use the Friedmann equation to substitute for $\dot{R}$. This will lead us to an expression that is also too general unless we make additional assumptions. During nearly all of its lifetime, the universe appears to have been matter-dominated. If we use this, we can repeat the arguments that led us to Eq. 2.52, although here there is an extra factor of R in the denominator. Calling $x = R(\tau')/R(\tau)$, we have, in this matter-dominated regime,

$$c \int_{\tau_i}^{\tau} \frac{d\tau'}{R(\tau')} = \frac{c}{R(\tau)H(\tau)} \int_{\frac{R(\tau_i}{R(\tau)}}^{1} dx \frac{1}{[x^2(1-\Omega) + x\Omega]^{1/2}} \tag{8.32}$$

Here $H(\tau) = \dot{R}(\tau)/R(\tau)$ and $\Omega = \rho(\tau)/\rho_c$. The integral can be done, producing

$$c \int_{\tau_i}^{\tau} \frac{d\tau'}{R(\tau')} = \frac{c}{R(\tau)H(\tau)(\Omega-1)^{1/2}} \tan^{-1}\left[\frac{\Omega-2}{2(\Omega-1)^{1/2}}\right]$$
$$- \tan^{-1}\left\{\frac{2\chi(\Omega-1)-\Omega}{2(\Omega-1)^{1/2}}[\chi\Omega - \chi^2(\Omega-1)]\right\} \tag{8.33}$$

where we have called $R(\tau_i)/R(\tau)$, χ. If we take the initial τ_i to correspond to the beginning of time, with $\tau_i = 0$ and $R(\tau_i) = 0$, we have the much simpler result

$$c \int_{0}^{\tau} \frac{d\tau'}{R(\tau')} = \frac{2c}{R(\tau)H(\tau)(\Omega-1)^{1/2}} \tan^{-1}(\Omega-1)^{1/2} \tag{8.34}$$

From these equations one can determine the event and particle horizons for any given model.

Here, as elsewhere, the discussion simplifies enormously if we take $\Omega = 1$. There are two reasons for this. In the first place, as we have noted before and will have occasion to note again, the choice $\Omega = 1$ is uniquely stable. Once Ω is 1, it remains 1. In contrast, let us suppose that Ω is only approximately 1 now and that we are matter-dominated. (It doesn't matter for the purposes of this argument whether we include in ρ such things as hypothetically massive neutrinos.) Then, $\Omega \simeq \rho_M/\rho_c$. But ρ_M scales as $1/R^3 \sim 1/\tau^2$ in the matter-dominated regime. Hence Ω will go rapidly to zero in the future. Only if Ω equals exactly 1 will it remain fixed at that value.

This aside, all the integrals become transparent when $\Omega = 1$, something dear to a theorist's heart. We have, for example, from Eq. 8.30, taking $\Omega = 1$ or, equivalently, $k = 0$,

$$r_h = \lim_{\tau \to \infty} \frac{2c}{R(\tau)H(\tau)}\left[1 - \sqrt{\frac{R(\tau_i)}{R(\tau)}}\right] \tag{8.35}$$

But, in the matter-dominated regime, for $\Omega = 1$, we have $H \sim \rho^{1/2} \sim 1/R^{3/2}$. Thus r_h is infinite, and so there is no event horizon.

It is interesting to work out the various distances with $\Omega = 1$. In general, the proper distance to a comoving coordinate r at time τ is given by

$$d(r)_P = R(\tau)\frac{\sin^{-1}\sqrt{k} \times r}{\sqrt{k}} \tag{8.36}$$

which means that, for $\Omega = 1$, we have $d(r)_P = R(\tau)r$. The luminosity distance d_L is given in general by $d_L = rR(\tau_0) \times R(\tau_0)/R(\tau_e)$, which implies that for $\Omega = 1$ we have

$$d_L = d_P \times \frac{R(\tau_0)}{R(\tau_e)} \tag{8.37}$$

so the two distances approach each other as the time interval between the emission and the absorption of the radiation shrinks.

We next want to repeat the discussion leading to Eq. 8.24, which allowed us to replace r, but now in the matter-dominated regime for arbitrary Ω. We use the generalization of Eq. 8.21,

$$\int_{\tau_e}^{\tau_0} d\tau' \frac{c}{R(\tau')} = \frac{\sin^{-1}\sqrt{k} \times r}{\sqrt{k}} \tag{8.38}$$

where the left-hand integral is given by Eq. 8.33. We shall first work it out explicitly for $\Omega = 1$, matter dominance assumed. In the $\Omega = 1$ case we have therefore

$$r = \frac{2c}{R(\tau_0)H(\tau_0)}\left[1 - \sqrt{\frac{R(\tau_e)}{R(\tau_0)}}\right] \tag{8.39}$$

Thus, here

$$d_L = \frac{2c}{H(\tau_0)}\frac{R(\tau_0)}{R(\tau_e)}\left[1 - \sqrt{\frac{R(\tau_e)}{R(\tau_0)}}\right] \simeq \frac{c}{H(\tau_0)}\left(Z + \frac{Z^2}{4}\right) \tag{8.40}$$

where Z is the red-shift factor given by $Z = R(\tau_0)/R(\tau_e) - 1$, which we have expanded in. If we identify, to lowest order, $Z = v/c$, the nonrelativistic Doppler shift, then to order Z we have Hubble's law. It will be recalled that in Chap. 2 of the Macropaedia we derived this law in Euclidean space as an exact relation. We now see that when space is not Euclidean, there are corrections that set in at order Z^2. For a nonEuclidean space it is not correct to associate the red shift with a Doppler velocity in any obvious way. We need a cosmological model to tell us when this association is approximately correct and what the correct relationships are.

For arbitrary Ω, but small Z and r, we can expand the integral in Eq. 8.33 in powers of Z, noting that $\chi = 1/(Z+1)$. Thus

$$r \simeq \frac{c}{R(\tau_0)H(\tau_0)}\left(Z - Z^2 \times \frac{\Omega + 2}{4}\right) \tag{8.41}$$

Hence

$$d_L \simeq \frac{c}{H(\tau_0)}\left[Z + \frac{Z^2}{2}\left(1 - \frac{\Omega}{2}\right)\right] \tag{8.42}$$

which reduces to the previous answer when $\Omega = 1$. We leave as an exercise for the reader to carry through the same set of arguments for the radiation-dominated regime and for the angular distance. We see, once again, that for small Z and arbitrary Ω, we have the Hubble law with corrections.

At the end of Chap. 1 of the Macropaedia we posed a puzzle that goes under the name of Olber's paradox. We saw that if we had a static universe filled with stars, the night sky would be ablaze with light. We said that we would resolve this puzzle in the context of the Big Bang expanding universe, and it is now time to make good on this promise. We begin by defining two volume elements: dV_{cm}, the comoving volume, and dV_P, the proper volume. We see from the Robertson–Walker metric that

$$dV_{\text{cm}} = \frac{1}{(1 - kr^2)^{1/2}} r^2\, dr \sin\theta\, d\theta\, d\phi \tag{8.43}$$

while the proper volume, which scales as $R(\tau)^3$, is given by

$$dV_P = \frac{R(\tau)^3}{(1 - kr^2)^{1/2}} r^2\, dr \sin\theta\, d\theta\, d\phi \tag{8.44}$$

Following the work of Chap. 1, we want to introduce the number density of stars. We shall define $n(L, \tau)\, dL$ as the number per unit proper volume, at proper time τ, of stars with absolute luminosity L lying between L and $L + dL$. Thus the number of sources, dN, at time τ lying between r and $r + dr$ with luminosity between L and $L + dL$ is given by

$$dN(L, r) = 4\pi \frac{R(\tau)^3}{(1 - kr^2)^{1/2}} n(L, \tau) r^2\, dL\, dr \tag{8.45}$$

In resolving Olber's paradox, the crucial thing is that the universe had a beginning a finite number of years ago. We are eventually going to integrate over all τ from that beginning to the present. We want therefore to convert the integral in Eq. 8.45 from an integral over r to an integral over τ. To this end, we can use Eq. 8.14 to write

$$dN(L, \tau) = 4\pi R(\tau)^2 \times n(L, \tau) \times r(\tau)^2\, dLc\, d\tau \tag{8.46}$$

where $r(\tau)$ is determined implicitly by Eq. 8.38. What concerns us is not the absolute luminosity L but the areal power $P(\tau)$ at a comoving coordinate r. Using arguments that we have already been over, we find that

$$P(\tau) = \frac{1}{4\pi r^2(\tau)} \times L \times \frac{R(\tau)^2}{R(\tau_0)^4} \tag{8.47}$$

Here τ represents the time of emission, which we are going to integrate over, while τ_0 represents the time at which we are observing all the starlight that has been emitted in the past. In the Olber's paradox context, according to which the universe had no finite beginning in time, we would integrate τ from, say, $-\infty$ to the present. But in the Big Bang model we would start the integral at some finite time—the time, whatever it is, when the stars began to shine. Let us call this time τ_*. Thus the total areal power integrated over all the stars and all times is given by

$$P = \int_{\tau_*}^{\tau_0} c\, d\tau' \frac{R(\tau')^4}{R(\tau_0)^4} \int dL\, Ln(\tau', L) \tag{8.48}$$

where the L integral is taken over all of the finite number of stars. Because of the limitations on the ranges of the variables τ' and L, P is clearly finite. It is sometimes said that in the expanding universe Olber's paradox is resolved because the energies of the photons are red-shifted. We see from Eq. 8.48 that there is such an effect, given in terms of the ratios of the R's. But that is not really what makes the integrals finite. What makes the integrals finite is that the universe was created a finite time ago and now contains a finite number of things—including stars. If we knew nothing else except the fact that the night sky is dark, we would have a strong argument that the universe has existed for only a finite time. Next we turn to some problems that the model universe we have so far described—the so-called Standard Model—present. We will see that we have to go beyond the Standard Model.

9

Beyond the Standard Model

All the work we have done so far has been done in the context of what has become known as the Standard Model. While there may be some difference of opinion as to exactly what the Standard Model encompasses, some of its elements would be agreed on by everyone. In the first place, everyone would agree that the Standard Model includes a hot Big Bang origin of the universe followed by an evolution that can be described in terms of Einstein's general theory of relativity. Most people would include in the Standard Model the Friedmann equations and the Robertson–Walker metric. They would also include the kind of elementary particle physics described in Chap. 5 of the Macropaedia. Implicitly, they would include quantum mechanics, statistical mechanics, and elements of the quantum theory of fields. Some people would restrict the notion of the Standard Model to include all, and only, those particles that we actually have direct evidence for. In this spirit, massive, unstable neutrinos would go beyond the Standard Model. What is remarkable is that the Standard Model, even strictly construed, has been able to account for so much. That is what this book, up to now, has been all about. In this final chapter, however, we will present several phenomena that appear to require our going beyond the Standard Model. This work will take us to the frontier of our subject. We will not be able to present things with the same didactic rigor as before since much of this story is still being told. Indeed, some of what we write now will be out of date before it can be printed. Hopefully, the basic ideas will remain intact. We see, at present, the general outlines, but we are not sure how to fill in the details. In construction, this chapter will resemble the Micropaedia. It will consist of sections linked together by the common theme of going beyond the Standard Model. It will make this chapter very long, but this organization seems more coherent than simply presenting a number of shorter unconnected chapters. In a more advanced book, some of these sections could be expanded severalfold, making use of concepts and techniques that go beyond the intention of this one. Our presentation will provide an introduction to this literature. The first topic we want to consider is what has come to be known as inflation.

9.1 INFLATION

Inflation has its genesis in several "puzzles" having to do with the initial conditions of the universe. We have put "puzzle" in quotation marks because, as we shall see, these are puzzles in the psychology or philosophy of what constitutes a legitimate scientific explanation rather than an outright contradiction between the model and some observed fact. We shall present two of these puzzles. The first one is known as the "flatness puzzle", although, at first sight, it may not be clear what it has to do with flatness. We have seen in the previous chapters of this book that the quantity $\Omega_0 = \rho_0/\rho_{c0}$, where, just to remind ourselves, $\rho_c = H^2 \times 3/8\pi G$, must lie, in the present universe, in the range between about 0.01 and 6. We have also seen that, unless $\Omega_0 = 1$, Ω will vary rapidly as function of time. In the future, if $\Omega_0 < 1$, then Ω will go to zero, and if $\Omega_0 > 1$, then Ω will first decrease until the universe begins to contract and then it will increase to infinity as the Big Crunch is approached. From this point of view, Ω_0 being so close to 1 would be an accident of the present epoch with no special meaning. But what about the past? To understand what happened in the past, we recall the fundamental Friedmann equation

$$\frac{8\pi G}{3} \times \rho(\tau) R(\tau)^2 - \dot{R}(\tau)^2 = kc^2 \tag{9.1}$$

Since the right-hand side of Eq. 9.1 is independent of time, we may write, where the subscript 0 refers to the present, $H = \dot{R}/R$, and where the time variables will be understood,

$$\frac{8\pi G}{3} \times \rho - H^2 = \left(\frac{8\pi G}{3} \times \rho_0 - H_0^2\right)\frac{R_0^2}{R^2} \tag{9.2}$$

or, in terms of the critical density,

$$\rho - \rho_c = (\rho_0 - \rho_{c0})\frac{R_0^2}{R^2} \tag{9.3}$$

We can rewrite this in terms of Ω to learn that

$$\Omega = \frac{\Omega_0}{\Omega_0 + (1 - \Omega_0)\dfrac{\rho_0}{\rho} \times R_0^2/R^2} \tag{9.4}$$

But in the cases of interest in the Standard Model, $\rho \sim R^{-n}$, where n is a positive integer: 4 for radiation and 3 for pressureless matter. Thus

$$\Omega = \frac{\Omega_0}{\Omega_0 + (1 - \Omega_0)(R_0/R)^{2-n}} \tag{9.5}$$

It is this formula that is very useful for determining Ω for the past and future. Indeed, it is the formula we used implicitly to make the statements we made about the future behavior of Ω. But now, let us run it backward to, say, the crossover between matter dominance and radiation dominance. This occurs at a temperature of about 10^4 K. Thus, since $R \sim 1/T^{v}$ and $n = 3$, the second term in the denominator is reduced by a factor of 10^4. If we go back even farther, we enter the radiation-dominated epoch in which the second term in the denominator falls even more rapidly to 0. If we go back far enough, we can make Ω absurdly close to 1. We can look at this in the following way. Suppose, for example, Ω_0 were to turn out to be $\frac{1}{2}$.

And suppose at a very early epoch we wanted to adjust the initial conditions so that things turned out this way. Then we would have to adjust Ω to some initial number that was fantastically close to 1. We would have to fine-tune the initial number so as to produce the present answer. There is nothing wrong with this from a purely technical point of view, but it goes against what we believe should constitute a real scientific explanation. We would have replaced an enigma by a mystery—the mystery of the existence of such a finely tuned initial condition. There are two ways out of this dilemma that one might imagine. It might be that in the end we actually have $\Omega_0 = 1$. In that case we would try to find a theory that contained some kind of built-in principle that made this inevitable. The second possibility is that Ω_0 is not actually 1, although close to it. In this instance we would look for a scenario for the early universe that would produce such an Ω_0 irrespective of the initial conditions. We have already seen similar scenarios. For example, we know that the cosmic background radiation is sensibly in a blackbody distribution. To account for this, all we have to account for is that at the epoch of recombination, when the photons were released, they were in equilibrium with the electrons. It does not matter what they were doing initially since an equilibrium distribution loses all memory of how it got that way. The photon scenario fits nicely into the Standard Model. But it is the Standard Model that has led us to Eq. 9.5 and to our Ω dilemma. Hence to find a suitable scenario for Ω we must go beyond the Standard Model.

This conundrum usually goes under the name "flatness" puzzle. To understand why, we shall rewrite Eq. 9.1 in a slightly different way. Using Eq. 9.1, we have at once that

$$\Omega - 1 = \frac{kc^2}{R^2H^2} \tag{9.6}$$

We see that $\Omega = 1$ corresponds to $k = 0$. But in the context of the Robertson–Walker metric, $k = 0$ corresponds to a metric of the form

$$ds^2 = R(\tau)^2[dr^2 + r^2(d\theta^2 + \sin\theta^2\, d\phi^2)] - c^2\, d\tau^2 \tag{9.7}$$

For each τ, this is just the ordinary flat space metric of special relativity. Hence we can restate the $\Omega = 1$ dilemma by asking, Why is space so flat? We can state our conundrum in yet another way by again rewriting Eq. 9.1. Thus

$$\frac{8\pi G}{3} \times \rho - \frac{\dot{R}^2}{R^2} = \frac{kc^2}{R^2} \tag{9.8}$$

In the radiation-dominated regime, assuming $\Omega \simeq 1$ but is not exactly 1, we have $R \sim \sqrt{\tau}$. Thus $\rho \sim 1/\tau^2$ and $\dot{R}^2/R^2 \sim 1/\tau^2$, while $1/R^2 \sim 1/\tau$. Thus, as τ approaches the Planck time, the two terms on the left-hand side of the equation become much larger than the term on the right. To compensate, we must have a nearly exact cancelation of the two left-hand terms, which certainly seems peculiar. In any event, this argument shows that in the radiation-dominated regime we can operate, for all practical purposes, as if k were zero. We turn now to the second of our two conundrums, the "horizon puzzle".

To understand the horizon puzzle we must find the distance, in our curved and expanding space, that light can travel from a comoving coordinate, which we will take to be $r = 0$, at $\tau = 0$, to a comoving coordinate r_h at the recombination time τ_c. This is significant because two objects that are farther apart than this distance, at time τ_c, are not in causal contact with each other. They are outside each other's past light cone, which means that since signals from one to the other cannot travel faster than light, the two objects cannot have been in communication in the past. As we saw in the last chapter, the comoving coordinates r_h are determined by the condition

$$\int_0^{r_h} \frac{dr'}{(1 - kr'^2)^{1/2}} = c \int_0^{\tau_c} \frac{d\tau'}{R(\tau')} \tag{9.9}$$

Therefore, for $k = 0$, the horizon distance d_h, the one that is measured in centimeters (or what have you), is given by

$$d_h = R(\tau_c)r_h = R(\tau_c)c \int_0^{\tau_c} \frac{d\tau'}{R(\tau')} \tag{9.10}$$

During this part of the expansion we are in the radiation-dominated regime with $R(\tau) \sim \sqrt{\tau}$. Thus, doing the integral and using the fact that $\tau_c \simeq 10^{12}$ s, we have

$$d_h = 2c\tau_c \simeq 6 \times 10^{22} \text{ cm} \tag{9.11}$$

The factor of 2 comes from doing the integral.

As the universe expands, this distance grows. If we extrapolate to the present, to what distance d_{h0} does the distance d_h correspond to now? This is interesting because if we observe now and see two phenomena separated by a distance larger than d_{h_0} in the sky, we can be sure that, given the correctness of our assumptions about the Big Bang expansion, they cannot have had a causal connection at the time of recombination. To find d_{h_0}, we scale d_h by multiplying it by $R(\tau_0)/R(\tau_c) = (\tau_0/\tau_c)^{2/3}$. The two-thirds power here represents the fact that since τ_c, the universe has been matter-dominated. Thus $d_{h0} = 2c\tau_c(\tau_0/\tau_c)^{2/3} = 2c\tau_c^{1/3}\tau_0^{2/3}$. To what distance should we compare this? A natural distance to compare it to is the size of the "visible universe". This is the farthest that light can have traveled since the creation. If space were flat and not expanding, this size, which we call l_0, would simply be $c\tau_0$. To take the expansion into account—we will always for simplicity assume $k = 0$ in these estimates—we write

$$l_0 = R(\tau_0)c \int_{\tau_c}^{\tau_0} d\tau'/R(\tau')$$

and do the integral. We are ignoring here the relatively brief epoch before recombination. In doing the integral we may then assume matter dominance. With these assumptions we find that $l_0 \simeq 3c\tau_0$. Comparing d_{h0} to l_0, we see that $d_{h0}/l_0 \simeq \frac{2}{3}(\tau_c/\tau_0)^{1/3}$. If we take $\tau_c \simeq 10^5 y$ and $\tau_0 \simeq 10^{10} y$, we find that the ratio is about 1:100. In other words, the visible universe is, according to this estimate, made up of about 100 distinct *noncommunicating* causal distances, or about 1 million *noncommunicating* causal volumes. Figure 1 illustrates the situation.

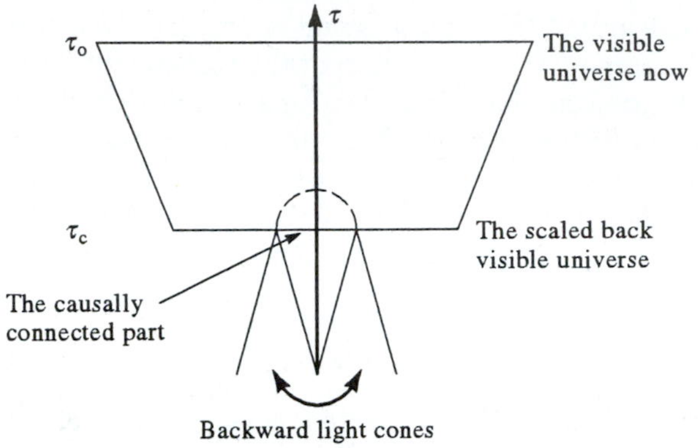

Figure 1 The visible universe and the causally connected particle horizon. We have drawn the nonintersecting backward light cones from the two causally disconnected end points.

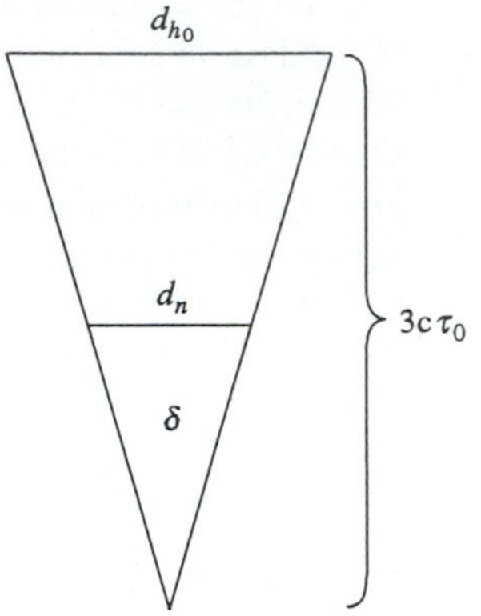

Figure 2 A triangle with the scaled proper particle horizon and the present size of the visible universe.

Before we comment on the significance of this result, we would like to state it in another way. Suppose, once again, that we take the causally connected horizon at τ_c and scale it forward to τ_0. What kind of angle on the sky would it cover? Astronomers usually discuss angular separations of celestial phenomena on the sky, and this will tell us what angle the scaled d_h will correspond to. Figure 2 illustrates the computation.

If the present size of the visible universe is taken as $3c\tau_0$, the scaled proper particle horizon as d_{h0}, and the angle of interest as δ, we have the relation

$$\delta = 2\sin^{-1}\left(\frac{d_{h0}}{3c\tau_0}\right) \tag{9.12}$$

Using this expression, we find that δ is about 1°. In other words, the Standard Model seems to tell us that any cosmological feature in the sky involving two points separated by an angle greater than 1° must have evolved from two points that were not causally connected at recombination! But how then are we to understand the isotropy of the cosmic background radiation? To give a quantitative notion of what it is we are to understand, in April 1992 the COBE satellite collaboration finally found, after people had been looking for years, evidence for an intrinsic nonisotropy in the cosmic background radiation. There has been evidence for several years for nonisotropy due to the motion of our galaxy, but in these new effects this galactic motion has been subtracted out. When we consider possible scenarios for structure formation, we will discuss this in more detail. The new data are taken at points separated by at least 10° all over the sky. The 10° has to do with the way the telescopes on the COBE are constructed and how the data are processed. This is well outside the limit for causally connected events. Yet the new data show that, at 10° or more, the nonisotropic effects are on the order of 6×10^{-6}. In other words, to 1 part in 100,000, the cosmic background radiation is isotropic at separations in the sky well outside the causal limit. To that level, the photon temperature appears to be the same in every direction. Yet, if the Standard Model calculation we have just given is right, none of these widely separated points could have been causally connected at recombination. How did they all know enough to be at the same temperature without simply putting this in as an arbitrary initial condition? Furthermore, this would have to be a different arbitrary initial condition than simply setting $\Omega = 1$. Indeed, even if it were to turn out that $\Omega = 1$, there would still be a horizon problem.

At the risk of beating this matter to death, we will state these two puzzles in a somewhat different manner that may point a possible way toward their resolution. At least it gives them a certain commonality. Our starting point is Eq. 9.6, which we write as

$$\Omega - 1 = \frac{kc^2}{R^2 \times \dot{R}^2/R^2} = \frac{3}{8\pi} \times \frac{k}{\hbar c} \times \frac{M_P^2 c^4}{R^2 \rho_c} \tag{9.13}$$

We have used the definition of the critical density and the Planck mass, that is, $G = \hbar c/M_P^2$. Now we can divide both sides of this equation by Ω to find

$$\frac{\Omega - 1}{\Omega} = \frac{3}{8\pi} \times \frac{k}{\hbar c} \times \frac{M_P^2 c^4}{R^2 \rho} \tag{9.14}$$

It is at this point that it is useful to bring entropy into the discussion.

Let us, as in Chap. 4 of the Macropaedia, call the entropy density s and the total entropy S, with the connection

$$S = R^3 s \tag{9.15}$$

or

$$R = \left(\frac{S}{s}\right)^{1/3} \tag{9.16}$$

Hence we can write Eq. 9.14 as

$$\frac{\Omega - 1}{\Omega} = \frac{3}{8\pi} \times \frac{k}{\hbar c} \times \frac{M_P^2 c^4}{\rho} \left(\frac{s}{S}\right)^{2/3} \tag{9.17}$$

Whatever scenario we are finally going to come up with will presumably involve events in the *very* early universe, since the Standard Model works wonderfully well for the "mere" early universe—microseconds after the Big Bang. In fact, we will be going all the way back to an era in which there are only quarks, the photonlike gluons that interact with them, and photons and leptons. Everything will be effectively massless. Thus the density ρ will take the form

$$\rho = N(T) \frac{\pi^2}{30} \frac{1}{(\hbar c)^3} (k_b T)^4 \tag{9.18}$$

Here we use k_b for the Boltzmann constant to avoid confusion with the curvature k. The number of degrees of freedom $N(T)$, which is, as indicated, a function of the temperature, is given by

$$N(T) = N(T)_{\text{BE}} + \tfrac{7}{8}N(T)_{\text{FD}} \tag{9.19}$$

where we have split $N(T)$ into its Bose–Einstein and Fermi–Dirac parts. To take the next step, we recall from Chap. 4 of the Macropaedia (Eq. 4.41) that for massless particles with no chemical potentials, ρ and the entropy density s are related by

$$s = \frac{4}{3} \times \frac{\rho}{T} \tag{9.20}$$

Putting everything together and evaluating the constants, we arrive at

$$\frac{\Omega - 1}{\Omega} = \frac{0.23k}{N(T)^{1/3}} \left(\frac{M_P c^2}{k_b T} \right)^2 \left(\frac{k_b}{S} \right)^{2/3} \tag{9.21}$$

We shall make good use of this formula shortly, but first we want to re-express the horizon problem in a similar way using entropy.

In this very early universe regime, before the particles have acquired masses and have begun to annihilate, the clock is given by the integral in Eq. 6.22 of the Macropaedia, namely,

$$\tau = \int_{T(\tau)}^{\infty} \frac{dT'}{T'} \left[\frac{3M_P c}{8\pi\hbar\rho(T')} \right]^{1/2} \tag{9.22}$$

This expression is derived under the assumption that $k = 0$, which is an excellent approximation in this epoch. Here the quantity $N(T)$ is sensibly independent of the temperature since the quarks and the other massless particles simply stick around. We can take $N(T)$ out of the integral and find that the causal length d_h at time τ is given in terms of the temperature, during the radiation-dominated regime, approximately by

$$d_h = 2c\tau_c = 1.21 \times \frac{\hbar c}{N(T_c)^{1/2}} \times \frac{M_P c^2}{(k_b T_c)^2} \simeq \hbar c \frac{M_P c^2}{(k_b T_c)^2} \tag{9.23}$$

This relation, incidentally, has implicitly assumed the conservation of entropy since in deriving it we take $R \sim 1/T$. The last approximate expression takes advantage of the fact that $N(T_c)^{1/2} \simeq 1$.

To write the ratio of the causal length to the scaled-back present length, we note that the scaled-back length is given by $c\tau_c = c\tau_0 \times T_0/T_c$. This again has implicitly assumed the conservation of entropy, that is, $R \sim 1/T$. Thus, ignoring all constants of order 1, we have the approximate relation

$$\frac{d_h}{l_0} \simeq \frac{M_P c^2}{k_b T_c} \times \frac{\hbar c}{l_0 k_b T_0} \tag{9.24}$$

But the dimensionless entropy within what we have been calling the visible universe, sometimes referred to as the "horizon entropy"—which we shall call S_H/k_b—is given approximately by

$$\frac{S_H}{k_b} \simeq l_0^3 \left(\frac{k_b T_0}{\hbar c} \right)^3 \simeq 10^{87} \qquad (9.25)$$

using the massless photons to compute the horizon entropy.

Therefore the last factor in Eq. 9.24 is, apart from constants, just $S_H/k_b^{-1/3} \simeq 10^{-29}$. From this point of view, we see that the horizon problem has at its base entropy conservation and the enormous present horizon entropy. But this is, as we will now see, also true of the flatness problem. If we suppose, as we have been doing in the Standard Model, that entropy is conserved as the universe expands, so that S is a constant, we can then evaluate it using data from the present epoch. This involves a subtlety. The quantity S we have in Eq. 9.21 is defined to be $R^3 s$. It is usually called the comoving entropy. We cannot evaluate it directly since we don't know the scale of R. But note that, ignoring all constants of order 1 and using Eq. 9.21 and the conservation of entropy, we can write $(k_b T_0/M_P c^2)^2(\Omega_0 - 1)/\Omega_0 \simeq k\,(k_b/S)^{2/3}$. We could, if we liked, include the factor of the curvature k in the definition of S, or we could rescale things so that $|k| = 1$. In any event, as the above expression shows, the quantity $k\,(k_b/S)^{2/3}$ is measurable in terms of Ω_0 and T_0. Indeed, if we take $|\Omega_0 - 1|/\Omega_0 \simeq 1$, we see that $(S/k_b)^{2/3}/k \simeq 10^{66}$. If we feed this huge number back into Eq. 9.21, we have the flatness problem re-expressed in terms of the entropy. Hence, from this point of view, to solve our problems we must find a way of violating entropy conservation, otherwise, the huge dimensionless entropy will cause trouble.

Since the Standard Model, which relies on entropy conservation, works so well for temperatures from at least several megaelectronvolts equivalent to the present, whatever entropy violation we invoke must have taken place even earlier. Particle theorists tell us that the interesting temperature regime for us to consider is $k_b T \simeq 10^{14}$ GeV. This is the temperature at which, according to theorists, all the interactions—gravity aside—acquire the same effective strength. Above this temperature the interactions are, we believe, unified into some symmetry larger than the ones we observe now. At this "unification temperature" of 10^{14} GeV equivalent, $M_P c^2/k_b T \simeq 10^5$. An examination of Eq. 9.23 shows us that this temperature is achieved about 10^{-35} s after the Big Bang, so we are talking about the very early universe indeed! We can now see how our two puzzles present themselves at 10^{-35} s. Putting the numbers into Eq. 9.21, we see that in this epoch $(\Omega - 1)/\Omega \simeq 10^{-49}$, a very strange initial condition. Using Eq. 9.24 at the unification temperature rather than T_c, we can also estimate the ratio of the causal to the scaled-back distance, which we find at this epoch to be about 10^{-25}. These numbers put the puzzles in sharp focus. As we have mentioned above, underlying them is the assumption that entropy is conserved. Suppose we give up this assumption. Suppose at $k_b T \simeq 10^{14}$ GeV a Big Event took place that generated entropy. How much entropy would have to be generated so that we could start off with a sensible initial condition and end up with both a causally connected universe and $\Omega = 1$? Let us call $S_{\langle}$ the entropy before the event took place and $S_{\rangle}$ the entropy after it took place. Let us introduce the entropy-boosting factor, which we call $\mathscr{G}$ in honor of A. Guth who was the first person to look at things this way, by the relation

$$S_{\rangle} = \mathscr{G}^3 S_{\langle} \tag{9.26}$$

To determine $\mathscr{G}$ we will begin our scenario at the Planck temperature $k_b T_P = M_P c^2 \simeq 10^{19}$ GeV. This is the temperature at which the initial conditions are to be set. We want both $(\Omega - 1)/\Omega$ and $2\tau/\tau_0$ to be about 1 initially. From the previous discussion, we see that both conditions can be met with $\mathscr{G} \simeq 10^{29}$. With this $\mathscr{G}$ we can have $S_{\langle}$'s that are of order 1, which is a pleasing initial condition. If we can find a scenario that produces a $\mathscr{G}$ of order 10^{29}, we seem to have resolved our problem. At least this appears to be a necessary condition. We now turn to the most popular scenario for producing such a $\mathscr{G}$, namely, inflation. We will also explore the question of whether entropy production alone is sufficient, as well as necessary, to resolve these puzzles.

We will approach this subject by way of a historical detour. As we mentioned in the introduction, when Einstein wrote his 1917 paper on the cosmological consequences of general relativity, he had available to him very little in the way of empirical evidence. It was not yet established that there are galaxies outside our own Milky Way, to say nothing of the fact that the universe is expanding. Einstein realized that such a static situation created difficulties for the general theory of relativity—the same difficulties Newton had come across when he tried to make a cosmology within a finite region of space. Such a model, because of the gravitational attraction of its component parts, will lead in general to a situation in which all the matter will end up clumping somewhere, which bears no resemblance to the essentially uniform distribution we observe. Newton resolved this problem by making space infinite so that there was no preferred place for the matter to collapse to, while Einstein resolved it by changing Newton's law of gravitation. In Newtonian gravitation, the gravitational potential energy $\phi(x, y, z)$ is determined by Poisson's equation

$$\nabla^2 \phi = 4\pi G \rho \tag{9.27}$$

where ρ is the matter density. Newton's law is then

$$\mathbf{F} = -\nabla \phi = m \frac{d^2}{d\tau^2} \mathbf{r} \tag{9.28}$$

In the Newtonian theory, one cannot have zero acceleration without having $\rho = 0$. As Einstein realized, cosmologically speaking, $\rho = 0$ is a poor approximation to what we observe. To reconcile a nonvanishing ρ with his static universe Einstein introduced a modified Poisson equation which he wrote in the form

$$\nabla^2 \phi + \Lambda = 4\pi G \rho \tag{9.29}$$

Here Λ is the notorious "cosmological constant" which we see has the dimension, as defined, of inverse time squared. In Einstein's static universe ρ is a constant, say

ρ_0, and we can reconcile its nonvanishing with his static universe by fixing the cosmological constant to have the value

$$\Lambda = 4\pi\rho_0 G \tag{9.30}$$

In this very special case, having $\phi = 0$ is compatible with a nonvanishing matter density. This new term can be adjoined to the equations of general relativity by adding to them a term of the form Λg_{uv}. This constitutes a "mutilation" of the original theory and also introduces a new parameter Λ of unknown origin. After Friedmann's work, and especially after Hubble's discovery that the universe does expand, the cosmological constant term fell into disgrace. Even Einstein disowned it, calling it his worst scientific "blunder". In 1923, Einstein wrote in a letter to the mathematician Hermann Weyl, "If there is no quasi-static world, then away with the cosmological term!" Nonetheless people like de Sitter and Lemaître placed the cosmological constant term in the context of Friedmann's expanding universe. In particular, in his 1932 paper, Lemaître rewrote the Friedmann equations to include such a term and, indeed, solved them in special cases. That is what we are going to do.

Let us begin with the matter-dominated case, Lemaître's example. We will assume that the total mass M remains constant and is related to ρ by $M = 4\pi/3 \times R^3\rho$. Lemaître introduced a modified Newton's law by adding a long-range repulsive force. The equation that results, which we give below, also follows from general relativity. It is

$$\frac{d^2}{d\tau^2}R = -\frac{4\pi}{3}\rho G R + \frac{\Lambda}{3} \times R = -\frac{MG}{R^2} + \frac{\Lambda}{3} \times R \tag{9.31}$$

The factor of 3 dividing the Λ, whose dimension is s^{-2}, is purely conventional. With this choice we achieve Einstein's static universe as a special case if we invoke Eq. 9.30. We can now multiply both sides of Eq. 9.31 by $\dot{R}$ and integrate to find

$$\frac{\dot{R}^2}{R^2} = \frac{8\pi}{3} \times G\rho + \frac{\Lambda}{3} - \frac{c^2 k}{R^2} \tag{9.32}$$

This is the modified Friedmann equation in question, which we rewrite as

$$\frac{\dot{R}^2}{R^2} = \frac{8\pi}{3} \times G\rho_{\text{eff}} - \frac{c^2 k}{R^2} \tag{9.33}$$

where

$$\rho_{\text{eff}} = \rho + \frac{\Lambda}{8\pi G} \doteq \rho + \rho_\Lambda \tag{9.34}$$

By definition, $\rho_\Lambda = \Lambda/8\pi G$. We may now ask, How big can ρ_Λ be in the present universe? In the present universe the k term in the Friedmann equation is certainly small and might even be zero. If we neglect it, we see that $\Lambda \lesssim \dot{R}^2/R^2$. Of course, Λ could be zero. However, if Λ were on the order of $\dot{R}^2/R^2$, then $\Omega_\Lambda = \rho_\Lambda/\rho_c$ would be of order 1, and the cosmological constant term could have a decisive effect on the expansion of the universe.

Our problem is that we do not have an underlying theory that fixes the cosmological constant. In fact, things are even more unsettled than Einstein believed. Even if we arbitrarily banish the cosmological constant from the equations of classical relativity, it can still come back to haunt us quantum mechanically. To see how this works requires a brief excursion into the quantum theory of fields. Most of the theories of elementary particles we deal with have in them at least one neutral scalar particle represented by a quantum field $\phi(\mathbf{r}, \tau)$. This field has dimensions of inverse length. If it is free and has a mass m, then there is associated with it a Hamiltonian density $\mathscr{H} = \hbar c/2\{(\dot{\phi}/c)^2 + (\nabla\phi)^2 + m^2c^2/\hbar^2 \times \phi^2\}$. If we integrate this density over space, we have the Hamiltonian H. We can find the average energy of any state of the Hamiltonian by sandwiching H between the state vectors. There will be a state of lowest energy, which is usually called the vacuum, although there can be a lot of activity in the quantum mechanical vacuum. Particles are continually created and annihilated in the vacuum—a sort of virtual reality. Let us call the vacuum state $|0\rangle$. Then the average vacuum energy density ρ_V is given by $\langle 0|H/V|0\rangle$. If we put the field ϕ into momentum space and work out the quantum mechanics, we find that the average vacuum energy density ρ_V of the free scalar field of mass m is given by

$$\begin{aligned}\rho_V &= \int \frac{d^3p}{(2\pi\hbar)^3} \times \frac{\sqrt{p^2c^2 + m^2c^4}}{2} \\ &= \frac{1}{4\pi^2}\int_0^{p_{\max}c} dp\frac{p^2}{\hbar^3} \times \sqrt{p^2c^2 + m^2c^4} \\ &\simeq \frac{1}{16\pi^2} \times p_{\max}c\left(\frac{p_{\max}}{\hbar}\right)^3 \end{aligned} \tag{9.35}$$

for $p_{\max}c \gg mc^2$. Clearly, ρ_V can have any value one likes depending on what one takes for $p_{\max}$. Now to the point. Gravity couples to any source of energy since, according to Einstein, energy is mass, give or take a factor of c^2. Thus the vacuum energy couples to gravity. This shows up as a term of the form $8\pi G\rho_V/c^2 \times g_{\mu\nu}$ which is added to the Einstein equations as a sort of virtual mass term. The extra factor of $1/c^2$ comes in because, as defined, ρ_V is an energy density while gravity couples to mass. But this is indistinguishable from a cosmological constant term. Indeed, it might be the only cosmological constant term since we do not know what the origin of any other such term might be. Different theorists have chosen different values of $p_{\max}$. A reasonable choice might, at first sight, be the Planck energy M_Pc^2 since one expects quantum mechanics to be affected once gravitation becomes strong. But this choice is disastrous since it leads to a ρ_V of about 10^{115} GeV cm^{-3}! This is to be compared to the critical density in the same units, $\rho_c = 1.054h^2 \times 10^{-5}$ GeV cm^{-3}. It is an outstanding problem to find a theory that deals sensibly with the quantum mechanically induced cosmological constant. Complicating matters, as we shall now proceed to discuss, is the fact that the inflationary scenario seems to demand a very early universe in which a nonvanishing cosmological constant played an essential role.

Although our primary concern will be with the very early universe, radiation-dominated epoch, it is instructive to solve the Friedmann equation with a non-vanishing cosmological constant in the matter-dominated regime. We will consider only the case $k=0$. The Friedmann equation can then be written as

$$\left(\frac{\dot{R}}{R}\right)^2 = \frac{\Lambda}{3}\left(\frac{\rho}{\rho_\Lambda}+1\right) \tag{9.36}$$

where, as before, $\rho_\Lambda = \Lambda/8\pi G$. If we introduce the present matter-dominated density ρ_0, then we have $\rho = \rho_0(R_0/R)^3$. If this is put into Eq. 9.36, the equation can be integrated. We leave it as an exercise to show that for $\Lambda > 0$—this would be the appropriate choice of sign of Λ if it is related to the average vacuum energy of a theory with a positive definite Hamiltonian—we have

$$\tau = \frac{2}{(3\Lambda)^{1/2}} \sinh^{-1}\left[\left(\frac{R}{R_0}\right)^{3/2} \times \frac{\sqrt{(\Lambda/3)}}{H_0}\right] \tag{9.37}$$

where, as usual, H_0 is the present Hubble parameter. Thus the lifetime of the universe, ignoring the first 100,000 yr, would be given by

$$\tau_0 = \frac{2}{(3\Lambda)^{1/2}} \sinh^{-1}\left[\frac{\sqrt{(\Lambda/3)}}{H_0}\right] \tag{9.38}$$

In obtaining these results, keep in mind that we are assuming that $k=0$ so that $\rho_0 = \rho_c = H_0^2 \times 3/8\pi G$.

To get a feeling, let us take $\Lambda = H_0^2$. In this case $\tau_0 = 0.95 \times 2/3 \times H_0^{-1}$. We have written the result this way because, it will be recalled, with no cosmological constant and the same assumptions, $\tau_0 = 2/3 \times H_0^{-1}$, which is the $\Lambda = 0$ limit of the right-hand side of Eq. 9.38. We bring all of this up because one could imagine finding oneself boxed in by experiments that show that $\Omega = 1$ and give an H_0 that produces an unacceptably short lifetime for the universe. We could compensate for this and still keep the nice features of the Friedmann universe by introducing some suitable Λ. One would then be confronted by the challenge of producing a theory that generates such a Λ.

We may now turn to the very early universe. Here ρ is dominated by radiation. One can also integrate Eq. 9.36. This is instructive since we have to use the conservation of entropy which we are assuming holds before the "event". Using the fact that $R^3 s = S$ and that $s = \frac{4}{3} \times \rho/T$, we find a relation between T and R. The R that occurs here has the dimensions of length—any convenient reference length. When we write the Friedmann equation, we will scale this length out. Thus

$$T = \left(\frac{S}{k_b}\right)^{1/3} \left[\frac{N(T)2\pi^2}{45}\right]^{-1/3} \frac{\hbar c}{k_b} \times \frac{1}{R} \tag{9.39}$$

We need this relation because we must write the Friedmann equation either in terms of T or R alone in order to solve it. Let us call the quantity

$$\epsilon = 7.90 \times \frac{G}{c^2}\left(\frac{S}{k_b}\right)^{4/3} \frac{\hbar c}{N(T)^{1/3}\Lambda}$$

It has dimensions of length to the fourth power, which we will scale out with R. The factor 7.90 comes from combining the various numerical constants. Recalling that the massless energy density ρ must be divided by c^2 to get the equivalent mass density, the Friedmann equation then takes the form

$$\frac{\dot{R}}{R} = \sqrt{\frac{\Lambda}{3}} \times \sqrt{\frac{\epsilon}{R^4} + 1} \tag{9.40}$$

The length dimension has scaled out, as advertised. If we integrate, we find

$$\tau = \frac{1}{2(\Lambda/3)^{1/2}} \sinh^{-1} \frac{R(\tau)^2}{\sqrt{\epsilon}} \tag{9.41}$$

or

$$R(\tau)^2 = \sqrt{\epsilon} \times \sinh 2\sqrt{\frac{\Lambda}{3}} \times \tau \tag{9.42}$$

If one plots R as a function of τ, one will find, as expected from the radiation dominance, that initially R increases as $\sqrt{\tau}$. However, when $2\sqrt{\Lambda/3} \times \tau$ becomes of order unity, we see that $R(\tau) \sim e^{\sqrt{\Lambda/3} \times \tau}$. There is an exponential growth of $R(\tau)$ with time. When this happens, one says that the universe has entered into a "de Sitter expansion". One of the models de Sitter considered amounted to solving Eq. 9.40, but setting $\epsilon = 0$. In other words, he imagined a universe with negligible matter density whose evolution was determined by the cosmological constant term alone. Under these conditions, he concluded, it would expand exponentially. It is curious that when Hubble published his 1929 paper, he thought that he might have discovered what he called the "de Sitter effect", that is, exponential inflation. The de Sitter expansion has some odd characteristics. In the first place, $\dot{R} = R_0\sqrt{\Lambda/3}e^{\sqrt{\Lambda/3} \times \tau}$, which means that the expansion velocity grows exponentially and will rapidly exceed the speed of light. This should not bother us since space is not a "thing" (Einstein), and there is no restriction on the overall expansion rate of space. However, this exponential expansion of space does show itself up as a peculiarity in the event horizon. The event horizon represents all the events that we will ever "see". A comoving event horizon coordinate r_E is given by

$$r_E = \lim_{\tau \to \infty} \int_0^\tau c\, d\tau'/R(\tau').$$

Let us suppose that the de Sitter expansion began at $\tau = 0$ with an initial R of R_0. Then, doing the integral, we find that

$$d_E = r_E R_0 = \lim_{\tau \to \infty} c\sqrt{\frac{3}{\Lambda}}(1 - e^{-\sqrt{\Lambda/3}\tau}) = c\sqrt{\frac{3}{\Lambda}} \tag{9.43}$$

What this means is that once inflation gets underway, there is an event horizon surrounding the point $r = 0$ with a radius of $c\sqrt{3/\Lambda}$. But the point $r = 0$ could be any point. Hence each point in the de Sitter space is surrounded by a sphere of radius $c\sqrt{3/\Lambda}$ that is isolated from the other spheres of the same radius surrounding the other points. Each sphere is a kind of private expanding universe isolated from the other private expanding universes.

Another curious feature in the de Sitter expansion has to do with the equation of state. So long as the temperature is above the temperature at which the nonadiabatic entropy-generating event occurs, entropy is conserved. This implies, as we saw in Chap. 4, that

$$dE + P\,dV = 0 \tag{9.44}$$

or

$$\frac{d}{d\tau}[R(\tau)^3\rho] + P\frac{d}{d\tau}R(\tau)^3 = 0 \tag{9.45}$$

But during the de Sitter expansion, $\rho = \rho_V$, where ρ_V is the constant vacuum energy density. For example, if there is an intrinsic cosmological constant, then $\rho_V/c^2 = \Lambda/8\pi G$. Thus the τ derivative of ρ_V vanishes, and we have

$$P = -\rho_V \tag{9.46}$$

In general, we expect ρ_V to be positive. Hence the de Sitter expansion is characterized by a negative pressure! This is unlike any normal liquid, for example. We can trace this quirk back to the fact that in the de Sitter expansion the cosmological constant energy increases as the volume increases:

$$E = R^3\rho_V = R_0^3 e^{\sqrt{3\Lambda}\tau} \tag{9.47}$$

To supply this increasing energy, work must be done on the system, hence the negative pressure. The total energy of the universe—the sum of the matter and the gravitational energy—is zero. To explain this would require a detour into quantum mechanical symmetries. But the basic idea here is that there is no absolute origin in time, just as there is no point in space that is singled out as the origin. This induces a symmetry called time and space translation invariance. The quantum mechanical quantities must depend on time and space differences and not on some absolute space-time coordinates. But it is the Hamiltonian of a system that generates displacements in time. To ensure that the system is time translation-invariant, it turns out that the Hamiltonian, when applied to a state describing the universe, must vanish. (It also implies that the total momentum must be zero.) Hence the total energy $\langle H \rangle$ must be zero. These are deep waters, however, this conveys the general idea. In the de Sitter epoch, the energy needed to run the expansion must be supplied by the gravitational field.

During the de Sitter epoch, the expansion is taken to be adiabatic. Thus, $T \sim 1/R$, which means that the temperature is falling exponentially as space is expanding exponentially. This means that energy—"latent heat"—is being stored in the vacuum. If one did not arrange for some sort of mechanism to stop the de Sitter expansion, it would go on forever. This would not correspond to our presently observed universe, which is not devoid of matter and appears not to be dominated by a cosmological constant. To avoid this, one speculates that the latent heat that is stored during inflation is released after the universe has inflated suitably. This released heat produces particles and radiation, and this dramatically increases the entropy. We will shortly discuss some of the speculative mechanisms that have been suggested for bringing this about, but first we want to run a few rough numbers. Let us suppose that the de Sitter expansion sets in at $k_b T = 10^{14}$ GeV, that is, at 10^{-35} s. This means that the visible universe then has a size of about 10^{-25} cm. Talk about tall trees from small acorns grow! At this time the entropy $S_{\langle}/k_b$ is, assuming it has been conserved since the initial condition, of order 1 compared to the 10^{87} we hope to end up with. This entropy $S_{\rangle}$ is related to $S_{\langle}$ by the equation $S_{\rangle}/S_{\langle} \simeq R^3_{\rangle}/R^3_{\langle} \times (T_{\rangle}/T_{\langle})^3$. We have used the approximate equality here since we are ignoring any possible changes in the numbers of degrees of freedom. In the standard inflationary scenario, reheating is supposed to return the temperature to about what it was before the inflationary expansion began, namely, $T_{\langle} \simeq T_{\rangle}$. Thus $S_{\rangle}/S_{\langle} \simeq e^{\sqrt{3\Lambda}\tau_{\rangle}}/e^{\sqrt{3\Lambda}\tau_{\langle}}$. During the de Sitter expansion the Hubble parameter H is given by $\sqrt{\Lambda/3} \simeq 1/\tau_{\langle}$. If we put this into the entropy relation, we have

$$\frac{S_{\rangle}}{S_{\langle}} \simeq \frac{e^{3\tau_{\rangle}/\tau_{\langle}}}{20} \tag{9.48}$$

If we solve for $\tau_{\rangle}$ using a ratio of $S_{\rangle}/S_{\langle} \simeq 10^{87}$, we find $\tau_{\rangle} \simeq 10^{-33}$ s. The universe has now expanded to about $e^{100} \simeq 10^{43}$ of its previous size. It is now a respectable 10^{18} cm. in radius. This is a factor of 10^{41} larger than the visible universe would have been at this time if it had been expanding in a peaceful radiation-dominated manner. That is a manifestation of the fact that, during the de Sitter epoch, the universe expands faster than the speed of light.

Wolfgang Pauli, who was an acerbic critic of scientific ideas, once criticized a grandiose scheme of a colleague, actually Heisenberg, by circulating a drawing which consisted of an empty space. Underneath Pauli had written the caption, "I can paint like Titian—only a few details are missing." That is the position we are in when it comes to realizing a concrete scenario that would implement these general ideas. Almost all cosmologists are in agreement that an inflationary phase followed by a nonadiabatic entropy gain would be a splendid way to solve the problems of the Standard Model. Indeed, as we will later argue, it may even be necessary. But how to get it to work? The schemes that have been proposed have one thing in common. They require a scalar field—perhaps several—whose energy densities for a brief time mock up a cosmological constant. To bring the exponential expansion to an end, one constructs the model so that the field decays. This decay into other particles and radiation is what produces entropy and reheats the universe. During the de Sitter expansion the other particles do not, it is assumed, contribute significantly to the energy density. They can be ignored, so that we are dealing, in essence, with the theory of a self-coupled scalar field ϕ, or possibly several such fields. The self-coupling implies that the scalar field Hamiltonian density $\mathscr{H}$ will contain, in addition to the kinetic energy terms, a term $\mathscr{V}(\phi)$—a potential energy density. We also allow for the possibility of a mass for the ϕ particle by including it, if required, in the $\mathscr{V}(\phi)$.

For simplicity we will suppose that there is only one kind of scalar meson. Then its Hamiltonian density can be written as

$$\mathscr{H} = \frac{\hbar c}{2}\left[\left(\frac{\partial}{\partial c\tau}\phi\right)^2 + (\nabla\phi)^2\right] + \mathscr{V}(\phi) \tag{9.49}$$

The scenarios then branch depending on what one takes for $\mathscr{V}(\phi)$. We shall present two of the most popular ones. The reader can judge how compelling they are.

In both scenarios one splits ϕ into its classical part ϕ_{cl}, and its quantum mechanical part ϕ_{qm}; that is, $\phi = \phi_{\text{cl}} + \phi_{\text{qm}}$. Here ϕ_{cl} is an ordinary numerical function of space and time—a classical field—while ϕ_{qm} is an operator. In what we are about to do, we shall ignore quantum mechanical corrections and write $\phi \simeq \phi_{\text{cl}}$. In the first scenario one takes $\mathscr{V}(\phi) = \lambda\hbar c\phi^4$. Written this way, with ϕ having dimensions of inverse length, the coupling constant λ is dimensionless. One could also take $\mathscr{V}$ to involve higher powers of ϕ, but this does not add anything fundamental to the discussion. (Taking $\mathscr{V} \sim \phi^2$ would amount to making ϕ a free field with a mass. This, it turns out, can also lead to inflation.) The resulting $\mathscr{H}$ is a theory that, because of its simplicity, has been studied for decades. There is even a learned school of theoretical physicists who claim that if it were to be solved exactly, it would be "trivial" in the sense that it would not permit any interesting physical processes, such as scattering, to take place. Cosmologists are a hardy lot and tend to ignore such complications. One should probably think of it as a convenient illustrative model. One then makes the assumption that at some time around 10^{-35} s, the classical field becomes, for a brief time, spatially homogeneous in a region of finite extent in space. These fields do undergo all sorts of fluctuations, so such a fluctuation might indeed occur. A weakness of the model is that this initial condition has to be assumed. It implies that in this region we have $\nabla\phi \simeq 0$, so this term can be dropped in the effective $\mathscr{H}$. What we wish to do next is to derive an equation that describes the evolution of ϕ as a function of τ. To do this in full detail would take us too deeply into the formalism of scalar field theory. However, we do not want to leave the impression that there is something mystical about it, so we will describe it briefly. The basic strategy is to use Eq. 9.45 in the form

$$\frac{d}{d\tau}(R^3\mathscr{H}) + P\frac{dR^3}{d\tau} = 0 \tag{9.50}$$

To apply Eq. 9.50, we must have an expression for P. This is the part that requires scalar field theory. In addition, it requires the assumption that this system evolves as a "perfect fluid". What this means is that if we construct the energy-momentum tensor density $\mathscr{T}_{\mu\nu}$ for the scalar field theory, the only non-vanishing components will be $\mathscr{T}_{00} = g_{00}\mathscr{H}$ and $\mathscr{T}_{ij} = Pg_{ij}$. This is essentially the definition of the pressure P as the coefficient of g_{ij} in the energy-momentum tensor. That there are no off-diagonal elements is really a consequence of the isotropy of space. Constructing the energy-momentum tensor is straightforward once we are given the Lagrange density $\mathscr{L}$ of the field theory. In the present case—a scalar field—the Lagrange density is related to the Hamiltonian density by the equation

$$\mathscr{L} = \hbar c\left(\frac{d}{dc\tau}\phi\right)^2 - \mathrm{H}.$$

If we drop, as we have been doing, the $\nabla\phi$ terms in the Hamiltonian density, we have simply $\mathscr{T}_{ij} = g_{ij}\mathscr{L}$. In our case, this implies that

$$P = \frac{\hbar c}{2}\left(\frac{d}{dc\tau}\phi\right)^2 - \mathscr{V}(\phi) \tag{9.51}$$

while

$$\mathscr{H} = \frac{\hbar c}{2}\left(\frac{d}{dc\tau}\phi\right)^2 + \mathscr{V}(\phi) \tag{9.52}$$

Note that the de Sitter epoch, which requires that $\mathscr{H} = -P$, can occur only when $(d/d\tau)\phi = 0$. The object of the present exercise is to see how this condition is approached and how it is departed from. To that end, we substitute Eq. 9.51 and 9.52 into Eq. 9.50 to find

$$\frac{d^2}{d\tau^2}\phi + 3\frac{\frac{d}{d\tau}R}{R} \times \frac{d}{d\tau}\phi = -\frac{c\,d\mathscr{V}}{\hbar\,d\phi} \tag{9.53}$$

where we have used $(d/d\tau)\mathscr{V} = (d/d\phi)\mathscr{V}\,(d\phi/d\tau)$. Equation 9.53 is a kind of "Newton's law" for the field ϕ, with an additional term that reflects the expanding universe. We are going to solve it in a self-consistent approximation. We will assume that $d\phi/d\tau$ is small and that $(d^2/d\tau^2)\phi$ is so small, compared to $d\phi/d\tau$, that it can be dropped. We will make this more precise shortly. We must then verify that our solution is consistent with our assumptions. If we take $\mathscr{V} = \lambda\hbar c\phi^4$, with no mass term and, to save writing, call $(d/d\tau)\phi = \dot{\phi}$, we have

$$\dot{\phi} = -\tfrac{4}{3} \times \frac{R}{\dot{R}} \times c^2\lambda\phi^3 \tag{9.54}$$

But we also have the Friedmann equation, which takes the form

$$\frac{\dot{R}}{R} = \frac{1}{M_P c}\left(\frac{8\pi}{3} \times \hbar c\right)^{1/2}\left(\frac{\hbar}{2c}\dot{\phi}^2 + \lambda\hbar c\phi^4\right)^{1/2} \simeq \frac{\hbar}{M_P}\left(\frac{8\pi\lambda}{3}\right)^{1/2}\phi^2 \tag{9.55}$$

where we have neglected the $\dot{\phi}^2$ kinetic energy term as compared to the potential energy. That is the sense in which we assume $\dot{\phi}$ to be "small". Putting these equations together, we have

$$\dot{\phi} = -\frac{M_P c^2}{\hbar}\sqrt{\frac{\lambda}{6\pi}} \times \phi \tag{9.56}$$

Solving Eq. 9.56, we find

$$\phi = \phi_0 \exp\left[-\left(\frac{M_P c^2}{\hbar}\sqrt{\frac{\lambda}{8\pi}}\right)\tau\right] \tag{9.57}$$

where ϕ_0 is a constant of integration.

In order for the de Sitter expansion to take over, we must have the ratio of the "kinetic" to the potential energy small; i.e., we must have $\mathscr{R} \doteq (\dot{\phi}^2 \hbar/2c)/(\lambda \hbar c \phi^4) < 1$. Inserting our solution, we find that

$$\mathscr{R} = \frac{1}{12\pi}\left(\frac{M_P c^2}{\hbar}\right)^2 \frac{1}{(c\phi_0)^2} \exp\left(\frac{2M_P c^2}{\hbar}\sqrt{\frac{\lambda}{6\pi}} \times \tau\right)$$
$$\simeq \frac{1}{12\pi}\left(\frac{M_P c^2}{\hbar}\right)^2 \frac{1}{(c\phi_0)^2}\left(1 + \frac{2M_P c^2}{\hbar}\sqrt{\frac{\lambda}{6\pi}} \times \tau\right) \quad (9.58)$$

We have expanded the exponential since if $\mathscr{R} \lesssim 1$, the exponent must be small. The question is, Can we find a set of plausible parameters that accomplishes this? Ultimately, if this scenario is to become more than amiable hand-waving, one would have to begin with the field theory, with a given coupling constant, and with the cosmology, and then deduce everything. At least we can see what would have to be deduced. Let us attempt to start the de Sitter expansion at $\tau = 10^{-35}$ s. Since all the particles are presumed massless, the only mass scale, or length, around is the Planck length. Hence it is hard to see what ϕ_0 could be except, approximately, the inverse Planck length $M_P c/\hbar$. These two requirements and our requirement on $\mathscr{R}$, as a little arithmetic shows, fix $\lambda \simeq 10^{-16}$. This is a very weak coupling indeed. It shows that particles that, even if they exist now, would have no direct experimental consequences could be decisive in determining the evolution of the early universe. We may next ask, Do we get enough inflation before our approximations break down? Clearly, as Eq. 9.57 shows, ϕ is tending exponentially to zero. This means that $\mathscr{V}(\phi)$ is going to zero rapidly, and that is what will shut the inflation off. Our criterion, derived from our entropy condition, is that $R(\tau_f)/R(\tau_i) \simeq 10^{29} \simeq e^{67}$. We can solve Eq. 9.55 setting $\phi \simeq \phi_0$. Thus

$$R(\tau) = R(\tau_0)\exp\left[\frac{\hbar}{M_P}\left(\frac{8\pi\lambda}{3}\right)^{1/2} \phi_0^2 \times \tau\right] = R(\tau_0)\exp\left[\frac{M_P c^2}{\hbar}\left(\frac{8\pi\lambda}{3}\right)^{1/2} \times \tau\right] \quad (9.59)$$

where, in the last expression, we have used $\phi_0 = M_P c/\hbar$. Equating the ratio of the R's with e^{67}, we find that we can do the job if inflation lasts until $\tau \simeq 10^{-34}$ s, which is consistent with our requirements on $\mathscr{R}$. This expansion by itself does not generate entropy. Indeed, one of the weaknesses of this model is that the reheating is left as a loose end. Perhaps after the de Sitter epoch the ϕ field acquires enhanced couplings to other fields and generates the entropy by the creation of particles. On the other hand, a very nice feature of the model is its handling of initial conditions. We have seen that a special kind of fluctuation is needed initially to generate the universe we observe. But fluctuations are going on all the time. Indeed, after the Big Bang, we can imagine that there were many different regions of space which at different instants were spatially homogeneous. These regions would inflate and produce universes of all kinds. Some would recollapse, and some would survive. None of them, by the argument we gave earlier, would be in communication with us. From this point of view our universe is a happy accident—a random fluctuation that went right.

Another weakness of this model, which generally goes under the rubric "chaotic inflation," is that the scalar fields are introduced ad hoc. As far as the model is concerned, their only task is to generate inflation. They do not appear to be connected to any other physics. We will now discuss another model of inflation that involves scalar particles that may actually have other uses. It also involves a very important theme in recent elementary particle physics that has the somewhat unfortunate name "spontaneous symmetry breaking." The conventional way of breaking symmetries in quantum mechanics—at least the one we learn first—involves adding to a Hamiltonian H, which respects some symmetry, an additional Hamiltonian H' that breaks it. If the symmetry is to be useful, the breaking term must be small and produce small changes in the wave functions. But there is another way of breaking symmetries that has come into use in the last few decades. In this so-called spontaneous symmetry breaking, the Hamiltonian retains its symmetry exactly, however, the solutions of the theory violate the symmetry. To take a completely elementary example of how the symmetry of the solutions can differ from that of the equations, consider the equation $d^2/dx^2\psi(x) = -k^2\psi(x)$, where k is a wave number. The Hamiltonian from which this equation could be derived has the symmetry $x \to -x$. This means that if $\psi(x)$ is a solution corresponding to the wave number k, so is $\psi(-x)$. However, we can find solutions that do not share the $x \to -x$ symmetry, namely, e^{ikx} and e^{-ikx}. We can also find a solution $\cos(kx)$ that does have the symmetry. In the field theoretic context, a commonly given example involves the scalar field Hamiltonian density

$$\mathscr{H} = \frac{\hbar c}{2}\left[\left(\frac{\partial}{\partial c\tau}\phi\right)^2 + (\nabla\phi)^2\right] + \frac{\lambda}{4} \times \hbar c\phi^4 - \frac{m^2c^4}{2\hbar c} \times \phi^2 \tag{9.60}$$

Despite appearances, the last term is not a conventional mass term because of the minus sign, although m has the dimensions of a mass. We now want to find the $\phi_{\rm cl}$ that minimizes this $\mathscr{H}$ which, incidentally, enjoys the symmetry $\phi \to -\phi$. Since the derivative terms are positive definite, we can do best by taking $\phi_{\rm cl}$ to be a constant in space and time. This leaves us with the task of minimizing the potential. That leads to the equation

$$\phi\left(\lambda\hbar c\phi^2 - \frac{m^2c^4}{\hbar c}\right) = 0 \tag{9.61}$$

This equation has three solutions: $\phi = 0$, $\pm(1/\sqrt{\lambda})(mc^2/\hbar c)$. We see that the value of the potential for $\phi = 0$ is zero, while the other two solutions produce the degenerate value of $\mathscr{V} = -(1/2\lambda)[m^4c^8/(\hbar c)^3]$. Figure 3 illustrates the situation.

We see that $\phi = 0$ is not a true minimum since the second derivative of the potential at that point is negative. Figure 3 also shows this clearly. This solution corresponds to what is called the "false vacuum." The true vacuua have $\phi = \pm 1/\sqrt{\lambda} \times mc^2/\hbar c$. But any given theory must choose one of these states as its vacuum. Once this is decided, then the $\phi \to -\phi$ symmetry is lost. The vacuum field is on one side of zero or the other. These solutions can also be used to generate masses for the scalar particles. If we define, for example, a new field χ by the equation

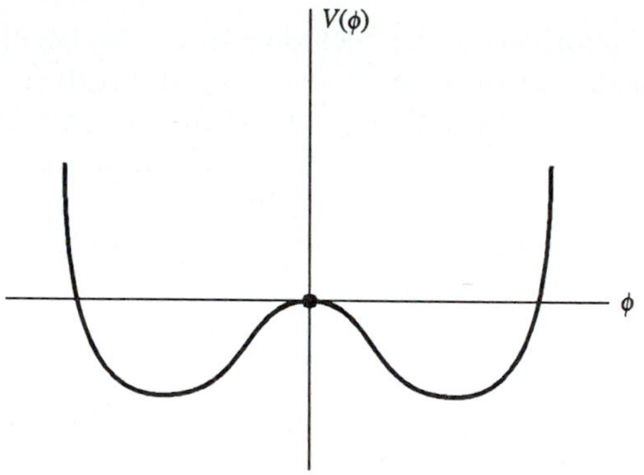

Figure 3 The potential $\mathscr{V}(\phi)$ with its minima.

$\chi = \phi + 1/\sqrt{\lambda} \times mc^2/\hbar c$ and rewrite $\mathscr{H}$ in terms of χ, we will find that the ϕ^4 part of the potential generates a term of the form $\frac{3}{2} \times m^2c^4/\hbar c$. This combines with the negative mass term to produce a positive mass term $2m^2$ for the χ field. The χ field also has all sorts of peculiar self-interactions. If we encountered the Hamiltonian density written in terms of it, we would simply say that it represents a massive scalar field with peculiar self-interactions. There would be no apparent trace of the original symmetry in this Hamiltonian density, although it is there behind the scenes. Most elementary particle physicists think that most, if not all, particle masses are generated this way. These scalar mesons are an example of what are called "Higgs mesons." (They are not the Higgs mesons that might have been discovered at the Supercollider, had it been built. Those Higgs mesons are supposed to have a mass energy of somewhere between 100 and 1,000 GeV. They are implicated in the unification of electromagnetism and the weak interactions.) These are the Higgs mesons that are supposed to break the electro-weak unification. On the other hand, in our cosmological scenario, the additional Higgs mesons that might be responsible for inflation have a mass energy of about 10^{14} GeV. They break the unification of the electro-weak force and the strong force. After they have done their work, the strong interactions are strong and the weak interactions weak.

How then does the inflationary scenario play itself out in this scheme? The scalar potential is written in terms of time- and hence temperature-dependent fields. This means that the shape of the potential changes with the temperature. Above a certain temperature that we can call T_c, the potential, if this idea is right, will look something like the $\lambda\phi^4$ potential. It will have one minimum, say $\phi = \phi_0$. At $T = T_c$ the potential will, one assumes, have evolved into a form something like the $\mathscr{V}(\phi)$ above, so that the ϕ field acquires a mass. We can then rewrite the Hamiltonian density in terms of something like the massive χ field. Now the field begins to evolve according to Eq. 9.53, where the mass term is included as part of $\mathscr{V}(\phi)$. Hence with a judicious choice of parameters we can have the right amount of inflation. Since the scalar field is massive, there is a new possibility, and this is one of the attractions of the scheme. The massive field, say χ, can be unstable and decay into other particles with a decay rate Γ. To describe this we would adjoin the term $\Gamma\dot{\chi}$ to Eq. 9.53. Of course, in a real theory this decay interaction would be included from the beginning in the underlying microphysics. For the scheme to work, we must adjust the parameters so that after the universe has sufficiently inflated, the $\dot{R}/R \times \chi$ term becomes less important than the $\Gamma\dot{\chi}$ term. When this happens, χ-particle decay takes over. This decay creates massive numbers of new particles since there were massive numbers of χ's, and these decay into two or more new particles. But increasing the number of particles increases the entropy. Indeed, the rate of change of the entropy is proportional to the rate of change in the number of particles. There is no entropy conservation here because the decay is out of equilibrium. As the temperature decreases, the daughter particles do not have enough kinetic energy to reconstitute their parents. However, the daughters can begin to interact among themselves and restore thermal equilibrium, meanwhile heating up the universe since the parent particles have dumped their mass energies into the kinetic energies of the daughters. The universe then continues to run its course. It is a pretty picture, and it may even be right. Specific models have been used to compute this scenario, and the results seem promising. What is lacking is a single compelling model that has the grandeur of what one is trying to explain. As Pauli said, ". . . only a few details are missing."

By this point the reader may share with the writer the feeling that a cosmology with a superluminal expansion epoch borders on science fiction. It is instructive therefore to present an argument as to why such an epoch may actually be *necessary* if, for example, we are to solve the horizon puzzle. Suppose, to the contrary, there is no de Sitter epoch but that for some unknown reason there is a jump in the dimensionless entropy at some temperature $T_{\rangle}$ to at least its present horizon value of 10^{87}—a Big Event. We may then estimate the temperature $T_{\langle}$ after this event occurs using the following reasoning. We assume that, after the event, the universe is radiation-dominated. All the energy that has gone into increasing the entropy has been dumped into radiation. This means, ignoring all constants of order 1, that $\dot{R}/R \simeq \hbar^{-1}(k_b T_{\langle})^2/M_P c^2$. This is the Friedmann equation for a radiation-dominated epoch. For this radiation, the entropy density s is given approximately by ρ/T. If we multiply this by the cube of the horizon distance d_h appropriate to this temperature, we get the total horizon entropy. In general,

$$d_h(\tau) = R(\tau)\int_0^{\tau} c\, d\tau'/R(\tau').$$

But, for both radiation and matter dominance, and even a mixture of the two, $d_h \simeq c \times R/\dot{R}$. Apart from constants of order 1, this is a good approximation to the integral. Indeed, for the argument that we are about to give to break down, we would have to have $cR/\dot{R}$ and d_h differ exponentially which, as we shall see, is just what happens in the inflation scenario. Putting things together, we have $S_H/k_B \simeq (M_P c^2/k_b T_{\langle})^3$, a relation we have seen before. If we put in 10^{87} for the dimensionless entropy and solve for $T_{\langle}$, we find $k_b T_{\langle} \simeq 10\,\text{eV}$! But this is a disaster. We have produced a very early universe characterized by a temperature appropriate to recombination, or lower. We might try to escape by saying that the Big Event occurred in the more recent universe when indeed $k_b T \simeq 10\,\text{eV}$, but then we would lose all the beautiful results of the Standard Model. How does the inflationary universe avoid this dilemma?

In the inflationary universe, the approximation $d_h \simeq cR/\dot{R}$ is totally wrong during the de Sitter epoch. Using our previous notation, during this epoch, $R(\tau) = R_0 e^{\sqrt{\Lambda/3}\tau}$. Thus $R/\dot{R} = \sqrt{3/\Lambda}$. To find d_h we suppose that the inflationary epoch began at some time τ_i and ended at some time τ_f. We may now do the integral to find that $d_h \simeq e^{\Delta\tau\sqrt{\Lambda/3}} \times c\sqrt{3/\Lambda}$, where $\Delta\tau = \tau_f - \tau_i$. The product $\Delta\tau\sqrt{\Lambda/3} \doteq N$, where N is the number of e foldings that occur during the inflationary epoch. Hence we can write $S_H/k_b \simeq e^{3N}(M_P c^2/k_b T)^3$. This leads to the condition $\ln(S_H/k_b)/3 = N + \ln(M_P c^2/k_b T_{\langle})$. If we take $S_H/k_b = 10^{87}$ and $k_b T = 10^{14}$ GeV, we find $N \simeq 55$. Fifty-five e foldings for the inflationary epoch are quite reasonable. In other words, we can have superluminal expansion followed by reheating to a reasonable temperature without having to give up the desirable features of the Standard Model. We are not forced to the absurd temperature that the noninflationary scenario led us to. We can also construct a complementary argument to suggest that one cannot really solve the flatness problem without invoking entropy production. Here we use Eq. 9.6,

$$\Omega - 1 = \frac{kc^2}{(\dot{R}/R)^2 \times R^2}$$

During the de Sitter epoch, $\dot{R}/R$ is constant, while R grows exponentially. Thus $(\Omega - 1)_{\text{bef}}/(\Omega - 1)_{\text{aft}} = e^{2N}$, where N is the number of exponential e foldings. Hence Ω is driven to 1, no matter what its initial value. But if this is done while conserving entropy, it is difficult to see how to avoid an exponential drop in the temperature. At the end of the process the universe would be left in a deep freeze out of which it would never emerge. Both entropy and superluminal expansion seem to be needed to solve the puzzles. With this introduction the reader will hopefully be prepared to tackle the huge and growing literature on inflation. We will come back to the role that inflation may play in the formation of structure in the early universe in the final section of this chapter, but we turn next to the problem of baryon production.

9.2 BARYON PRODUCTION

Before we state the problem, let us begin by re-expressing the photon entropy density s_γ in terms of the photon number density n_γ. We leave it as an exercise to show that these are related by

$$s_\gamma = 3.61 n_\gamma \tag{9.62}$$

This makes it clear that the conservation of the total entropy and the conservation of the total number are, in this instance, one and the same. We can make use of this relationship in defining a quantity $\mathscr{R}_B$ as

$$\mathscr{R}_B \doteq \frac{n_B - n_{\bar{B}}}{s_\gamma} \tag{9.63}$$

This quantity bears a family relationship to the quantity $\eta = n_B/n_\gamma$, defined in Chap. 6. Indeed, it would have made equal sense to have defined η in terms of the net baryon density $n_B - n_{\bar{B}}$ since, and this is the subject of the present section, $n_{\bar{B}}$ is sensibly zero. With this modification, $\mathscr{R}_B = 0.28\eta$. Since the total photon entropy in the universe has been a separately conserved quantity since the photons decoupled and since the net total baryon number is, as far as we know, a nearly, even if not exactly, conserved quantity, $\mathscr{R}_B$ has been nearly independent of time for most of the lifetime of the universe. The puzzle is why do the antibaryons make such a small contribution to $\mathscr{R}_B$? Where are the antiparticles?

Before we try to answer these questions, let us first discuss the evidence. It turns out that it is surprisingly difficult to prove that astronomical objects are not made of antimatter. The moon and the planets are of course an exception since they have been in contact with terrestrial matter which did not dissolve in a cloud of gamma radiation. To understand the difficulty in determining the matter or antimatter content of less tangible astronomical objects, we must back up and complete a discussion of the symmetries, obeyed and disobeyed by the various interactions, that we began in Chap. 5—the chapter on elementary particles. In that chapter we pointed out that a series of experiments performed in 1956 showed that parity symmetry P was violated by the weak interactions. But the same group of experiments also showed that a second symmetry known as "charge conjugation" C also broke down. This symmetry, which came into physics in the 1930s, was based on the intuition, which seemed to be borne out by experiment, that theories should be symmetric between particles and antiparticles. This led to selection rules. A simple example is the decay of the π^0. The π^0 is its own antiparticle and goes into itself under the operation of charge conjugation. The photon is also its own antiparticle, but its situation under charge conjugation is a bit more complicated. Charge conjugation changes the sign of the charge or, more generally, the sign of the charge-bearing current. But this current j_μ couples to the photon field A_μ with a coupling of the form

$$\sum_\mu j_\mu A_\mu.$$

Hence to keep the theory invariant, the photon field must go into its negative. This means that charge conjugation forbids the decay $\pi^0 \rightarrow \gamma + \gamma + \gamma$. Indeed, experiment shows that this decay, which has never been seen, has a branching ratio of less than about 10^{-8}, confirming the symmetry prediction. But since the weak interactions violate charge conjugation, there should be a tiny three-gamma branch of the π^0 decay—one that is too small to have been observed.

Until 1963, workers in this field lived under the happy illusion that, although C and P were separately violated, the product CP still remained a good symmetry. Part of this feeling had to do with what was known about the nature of the combined transformation CPT, where T stands for what is referred to as "time reversal." Unlike P and C, time reversal had a clear status in classical physics, in particular in statistical mechanics where it was related to the microreversability of collisions

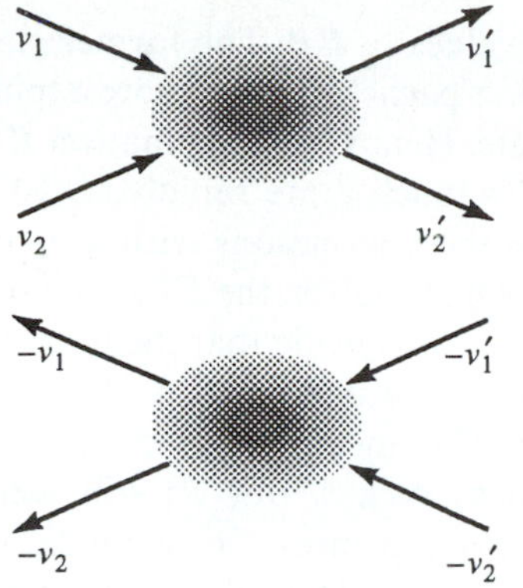

Figure 4 A collision, and below it, its microreversed counterpart.

among the molecules in a gas. In Fig. 4 we give an illustration of a two-particle collision and its microreversed counterpart with the velocities inverted.

The principle of the microreversability says, in essence, that if one of these collisions is a legitimate physical process, obeying all the conservation laws, then so is the other. This gave rise to the great debates among physicists and philosophers of science as to how a theory in which this principle is embedded could allow an "arrow of time," the flow of time in what appears to be one "direction." Most people agree that it is a matter of the initial conditions. If one chooses a highly improbable initial condition, then the system evolves in the direction of more and more probable configurations, even though all the collisions have their reversable counterparts. With the advent of quantum mechanics, the quantum mechanical analog of this symmetry was introduced. It turned out to be a quite subtle business since the operation that produced it had to be what is known as "antilinear." We can see this from the ordinary Schrödinger equation in which the complex number i and the time derivative $\partial/\partial t$ occur in the combination $i\partial/\partial t$. Hence if we reverse the sense of time, we must also change the sign of i to keep the theory invariant. It is this changing of the sign of i that is the antilinear part of the operation. The operation of CPT involves the application of all three of these symmetries. It turns out that it is very difficult to find a quantum theory that, at one and the same time, respects the theory of relativity and violates the CPT theorem. One can construct such theories, but they are very artificial and very ugly. The CPT symmetry is very powerful. In particular, it means that if CP is a good symmetry, then so is T. This is what was tested in 1963.

The system that was used was the same oscillating $K^0 - \bar{K}^0$ complex we discussed in Chap. 7. In the neutral K-meson system we take advantage of the fortunate fact that the two states that decay with definite lifetimes, the states we called K_S^0 and K_L^0, also have opposite symmetries under CP. In a world in which CP is a good symmetry, the relation between the K^0 and the $\bar{K}^0$ is given by $\bar{K}^0 = CPK^0$. Hence the linear combinations $K^0 \pm \bar{K}^0$ have opposite symmetries under CP. The K mesons have spin 0, so whatever state they decay into must also have zero total angular momentum. It is not very difficult to show that a final state—to take a relevant example—with two π^0's and a final state with three π^0's, both of which have zero total angular momentum, have opposite values of *CP*. The former state is even under *CP*, while the latter is odd. In general, a particle decays more rapidly into a two-particle state than into a three-particle state. Hence the combination $K^0 + \bar{K}^0$, which is even under *CP*, will decay into two π^0's much more rapidly, $\sim 10^{-10}$ s, than the combination $K^0 - \bar{K}^0$, which decays into three pi mesons with a lifetime of about 10^{-8} s. All of this assumes *CP* invariance which—given the *CPT* theorem—means *T* invariance. In 1963, the remarkable discovery was made that the long-lived neutral K^0 meson, K^0_L, which had been thought to be odd under *CP*, had a branch of about 1 in 1,000 of decays into two pi mesons. This meant that *CP* invariance, and presumably time reversal, was violated in these decays. But we still seem to have *CPT*. The *CPT* theorem has many striking consequences. For example, from it we can deduce that a particle and its antiparticle have identical masses and lifetimes. All our evidence is consistent with this. In fact, in view of the noninvariance of *CP*, we now define the antiparticle of a particle A, $\bar{A}$, as $\bar{A} \doteq (CPT)A$. The fact that *CPT* appears to be absolutely conserved does not mean that the two-pi zero decay described above is absolutely forbidden, since the final pion states do not have a simple symmetry property under *CPT*.

The absolute validity of *CPT* and the near validity of *CP* whose violation has never been observed except in the neutral *K* meson system, are what makes it so difficult to determine whether or not a distant astronomical object is made of antimatter. The energy levels of atoms and anti-atoms are sensibly the same. These levels are determined by the electromagnetic interactions, and these respect *CP*. This means that the spectral lines coming from an antistar would be indistinguishable from the spectral lines coming from a star. Indeed, all our evidence about the absence of antimatter in the universe is somewhat indirect. Here are a few examples. The sun emits a "solar wind" consisting of protons and helium nuclei. If there were significant amounts of antimatter anywhere in the Solar System, it would collide with the solar wind, and this would produce very energetic x-rays of a kind that have not been observed. Cosmic rays come from outside the Solar System. They consist in part of very energetic protons. These protons can collide with ordinary matter and produce antiprotons. The observed antiproton-to-proton ratio in cosmic rays is about 10^{-4}. This is entirely consistent with the absence of primary antiprotons. This is the kind of evidence that is usually cited to support the claim that there is little or no cosmological antimatter in the universe. But over and beyond this, there is the apparently insuperable problem of what would have happened if at some early epoch there had been as much matter as antimatter. What would have kept it from annihilating—the matter against the antimatter? No one has come up with a scenario that leaves enough matter behind to fit what we observe. One is forced to the conclusion that matter-antimatter asymmetry must be nearly as old as the universe itself.

We are once again faced with an issue in the philosophy of scientific explanation. We could, if we didn't care about a real explanation, simply assume that baryon number $n_B - n_{\bar{B}}$ was an absolutely conserved quantity and that its initial value was such that $n_{\bar{B}}$ was essentially zero. This is very unsatisfying to a scientist. If we want to produce what we would think of as a scientific explanation, one element must be the assumption that baryon number is not absolutely conserved. This is something that again takes us beyond what most people think of as the Standard Model. However, even before the cosmological question was actively considered, particle theorists were already trying to extend the Standard Model so as to unify the strong, weak, and electromagnetic interactions. These extensions involve marriages between baryons and leptons which allow interconversions between them. Because of these interconversions, neither the baryon number nor the lepton number, which measures the difference between leptons and antileptons, is absolutely conserved, although combinations like the difference between these numbers may be. This means, for example, that these theories predict that the proton can decay. As we mentioned in Chap. 7, the original motivation for building massive subterranean detectors was just to detect these decays. All sorts of decay modes have been looked for. For example, $p \rightarrow e^+ + \pi^0$ or $p \rightarrow \nu + \pi^+$. The net result of these investigations is that the proton lifetime must be longer than about 10^{31} yr! This has ruled out some of the simpler unified theories but not the idea. Hence, to explain the baryon number, we must assume that some such theory will eventually work.

The first person to have tried to adapt these ideas to an explanation of the baryon number seems to have been Andrei Sakharov, although his 1967 paper is both very qualitative and difficult to read. The key insights are two. The first is that while such baryon-violating decays are at the present time extremely weak (as is illustrated by the incredibly long proton lifetime), there could well have been an early universe epoch in which they were strong. The effective coupling constants of the various interactions—the actual coupling that determines their observable strength as opposed to some number one puts into a Lagrangian—depend on the energies involved, hence on the ambient temperature. Indeed, the idea of these unified theories is that there is a unification temperature at which all interactions have the same effective strength. At this temperature the baryon number nonconserving interactions might have been quite significant. The second important insight involves the notion of phase changes. In the universe we now inhabit, quarks seem to be unobservable as free particles. There is very strong evidence that the particles we do observe are made up of quarks, but they appear to be confined permanently. However, we believe that this was not always the case. We think that before some equivalent temperature—several theories suggest about 10^{14} GeV—the elementary particles we now observe did not exist. What did exist was an environment of free quarks, radiation, leptons, and, very likely, some unstable particles that are no longer around. In this very early universe, the problem of baryon number would take on quite a different appearance. The only "baryons" around would be the quarks and antiquarks. They would be about as numerous as photons. Hence for this epoch we may write

$$\frac{n_B - n_{\bar{B}}}{n_B} \simeq \frac{n_q - n_{\bar{q}}}{n_q} = 1 - \frac{n_{\bar{q}}}{n_q} \simeq 10^{-9} - 10^{-10}$$

or,

$$\frac{n_{\bar{q}}}{n_q} \simeq 1 - (10^{-9} - 10^{-10})$$

This means that a tiny difference in the quark-antiquark baryon number in this early epoch would account for the presently observed totally asymmetric baryon number, providing that baryon number has been nearly conserved between then and the present. We must now construct a plausible scenario.

The first question we must answer is, When is this baryon-producing process supposed to have taken place? Whatever else, it must have taken place after the entropy-producing postinflationary sequence of events. The reason is to be found in $\mathscr{R}_B$. If the entropy factor in the denominator jumped by a huge amount, the baryon number would drop essentially to zero. Any baryons produced before this epoch would become imperceptible. This dilution of particles by inflation is actually a very useful thing, at least for getting rid of some unwanted particles. We did not mention it above because it has a somewhat more theoretical character, but another of the puzzles of the Standard Model is the production in the early universe of particles that do not seem to be around at present. The most notorious case is that of the magnetic monopole. This is a free magnetic "charge," which is the analog of the free electric charge. In conventional electrodynamics there is no such animal, but the same unified theories that are used to describe the quark phase change also predict the production of a large number of magnetic monopoles before inflation. Despite an intense search and a few false alarms, there is no credible evidence that such charges exist at present. If they can be diluted away by inflation, that would be a very fine thing indeed.

The second point we should consider is the matter of equilibrium. We claim, and will explain, that so long as quarks and antiquarks are in equilibrium, there can be no net change in the baryon number. This can be demonstrated on a number of levels, the most sophisticated one being the actual study of the rate equations. Using the *CPT* theorem and the rate equations, one can give a detailed proof of this proposition. Here we want to do something somewhat more homespun. We have already argued that any change in the baryon number must involve an interaction that does not respect baryon number conservation. In such theories, we will generally have two-body scattering processes of the form $q + q \leftrightarrow \bar{q} + \bar{q}$, where the q's are any species of quark. These clearly violate baryon number conservation. Suppose that these processes have produced equilibrium. In that case the quark distributions would take the Fermi–Dirac form

$$f(q) = \frac{1}{\exp[(p^2c^2 + m^2c^4)^{1/2} + \mu_q]/kT + 1}$$

where μ_q is the chemical potential. The antiquark distribution would take the same form, with μ_q replaced by $-\mu_q$ since the chemical potentials of particles and antiparticles in equilibrium are equal and opposite. The *CPT* theorem assures us that the masses of the quarks and the antiquarks are identical. But we know that in equilibrium chemical potentials are conserved. Therefore the equilibrium two-body process produces the equation $\mu_q + \mu_q = \mu_{\bar{q}} + \mu_{\bar{q}} = -\mu_q - \mu_q$. Thus $\mu_q = 0$. But this means that the number densities n_q and $n_{\bar{q}}$ are equal. There is no net baryon number when the quarks are in equilibrium under the conditions we have described. Hence baryon production is an out-of-equilibrium phenomenon.

The scenario embodying these features that has received the most attention involves assuming that there is a very heavy, say 10^{15} GeV equivalent energy, meson, which is usually called the X particle. It no longer exists because it is unstable and has decayed away. This meson is assumed to have decay modes that violate C, CP, and baryon number conservation but respect CPT. In some versions, the $\bar{X}$ (the anti-X) is identical to the X, and in some versions it isn't. We will consider the version where they are distinct. In this version the X's are electrically charged and have the decay modes we write generically as $X \rightarrow q + q$ and $X \rightarrow \bar{q} + \bar{l}$, where l is some antilepton. To be specific, let us assign a charge $\frac{1}{3}$ to the X and assume that it decays via $X \rightarrow u + d$ and $X \rightarrow \bar{u} + e^+$, where u is the up quark and d is the down quark. The $\bar{X}$ then decays via $\bar{X} \rightarrow \bar{u} + \bar{d}$ and $\bar{X} \rightarrow u + e^-$. Since the X is a meson with zero baryon number and the quarks have baryon number $\frac{1}{3}$, while the leptons have no baryon number, these decays clearly violate baryon number conservation. We can also use the X to induce two-body scattering processes such as $u + d \rightarrow \bar{u} + e^+$. An important feature of this specific model is that although baryon number and lepton number are not separately conserved, the quantity $B - L$ is conserved. This means that in equilibrium there is a nonvanishing chemical potential corresponding to this conservation law. For example, the quark-scattering process described above leads to the equation $3\mu_q = -\mu_l$. This suggests an interesting possibility. If initially we had $B = L$, then this relation would be preserved in time. In this model any process that generated B would also generate L. As the temperature drops so that the equivalent ambient energy is less than the rest mass energy of the X mesons, the decays will fall out of equilibrium with their inverse reactions. That is the epoch when baryon number will be produced. To make things slightly more quantitative, let us call the width of the X, Γ_X—the inverse of Γ_X is proportional to the X lifetime—and let us call the width of the $\bar{X}$, $\Gamma_{\bar{X}}$. The CPT theorem then implies that $\Gamma_X = \Gamma_{\bar{X}}$. Since there are two decay modes for the X, we can write $\Gamma_X = \Gamma_{qq} + \Gamma_{\bar{q}\bar{l}}$. It is customary to make the definitions

$$r \doteq \frac{\Gamma_{qq}}{\Gamma_X} \tag{9.64a}$$

and

$$1 - r = \frac{\Gamma_{\bar{q}\bar{l}}}{\Gamma_X} \tag{9.64b}$$

We also define

$$\bar{r} \doteq \frac{\Gamma_{\bar{q}\bar{q}}}{\Gamma_X} \tag{9.65a}$$

and

$$1 - \bar{r} = \frac{\Gamma_{ql}}{\Gamma_X} \tag{9.65b}$$

where we have used the fact that $\Gamma_X = \Gamma_{\bar{X}}$. The *CPT* theorem does not require that $\bar{r} = r$, since these refer to a single decay mode. Suppose a pair of X and $\bar{X}$ mesons decay. What is the baryon number produced per pair of such decays? We will consider the specific example where the X decays into a u and a d, with a net baryon number of $\frac{2}{3}$, and into a $\bar{u}$ and an e^+, with a net baryon number of $-\frac{1}{3}$. The antiquarks all haved baryon numbers of $-\frac{1}{3}$. The $\bar{X}$ produces final states with the opposite baryon numbers. We may now compute $N_B(X, \bar{X})$—the net baryon number produced per $X, \bar{X}$—in terms of r and $\bar{r}$. Thus

$$N_B(X, \bar{X}) = \tfrac{2}{3}r - \tfrac{1}{3}(1 - r) - \tfrac{2}{3}\bar{r} + \tfrac{1}{3}(1 - \bar{r}) = r - \bar{r} \tag{9.66}$$

One can show that if *CP* is good, a more powerful condition than the *CPT* theorem, then, indeed, $r = \bar{r}$. This is plausible since in a world in which *CP* is good, we can define the $\bar{X}$ to be the *CP* conjugate of the X. In this world, the u is the *CP* conjugate of the $\bar{u}$, and e^- the conjugate of e^+. The two-body final states are simple enough so that they go into each other under *CP*, hence the result $r = \bar{r}$. But we know from the $K^0, \bar{K}^0$ system that *CP*-violating reactions do exist. We don't know why they exist, and it would be very satisfying to connect them to some deep cosmological consideration such as the production of baryons. We can get a crude idea of the magnitudes involved in the difference $r - \bar{r}$ as follows. In this picture, the net baryon number density $n_B - n_{\bar{B}}$ is given roughly by $n_B - n_{\bar{B}} \simeq n_X(r - \bar{r})$, where n_X is the X-meson density. It enters here since in this model the baryon production is proportional to the X decay. But there is no reason to think that during this epoch the X particles are any less numerous than the photons, so that $n_B - n_{\bar{B}} \simeq n_\gamma(r - \bar{r})$. To fit the presently observed data on $(n_B - n_{\bar{B}})/n_\gamma$, we must therefore have $r - \bar{r} \simeq 10^{-9} - 10^{-10}$. Without a theoretical model it is difficult to say whether this is a big number or a small number. It is probably fair to say that, as with inflation, the idea of producing baryon number by out-of-equilibrium decay processes seems right, but we don't have a compelling theory: "only a few details are missing."

9.3 STRUCTURE FORMATION

One of the most striking features of the universe, if one looks at it on scales that are not too large, is its lack of homogeneity. It is full of structure. There are galaxies, clusters of galaxies, and even superclusters. In addition, there are galaxies that seem to be distributed in filamentary lines, one of which is known to astronomers as the Great Wall—a row of galaxies that appear to block the sky. Much of this information comes from the NASA (Infrared Astronomical Satellite) (IRAS) which was launched in 1983 and has been wonderfully productive. It works in the 60- and 100-μ wavelength regime. This infrared radiation penetrates the obscuring dust in our own galaxy, allowing one to "see" features that would be otherwise unobservable. The problem immediately presents itself of how to reconcile all of this diverse structure with the Standard Model. The Standard Model gives a wonderful account of how, for example, blackbody radiation could have come about and why it should be so uniform. But what breaks this uniformity and produces the structure we observe? On this, the Standard Model is mute. It has been evident for decades that this structure formation must be the work of gravitation. Something must have perturbed the uniform background early in the history of the universe, allowing gravitation to do its work of clumping matter. But what? What produced the fluctuations and ensured that they were of just the right magnitude to allow for the evolution of what we observe but not so large as to conflict with the uniformity that we also observe. This is a problem in the philosophy of scientific explanation similar to the first two we have discussed. We could simply put these initial fluctuations in by hand, but then we would no longer be doing science. The subject has been given considerable currency by the discovery, in the spring of 1992, of tiny anisotropies in the cosmic background radiation. This is the work of the COBE satellite to which we have already referred. These fluctuations, which now appear to have been confirmed in experiments done both in balloons and from the ground, are the first evidence we have of the structure-forming mechanism at work. They have ushered in a new era of cosmological research which will keep cosmologists busy for decades.

As a preliminary to the discussion of these matters we return to a subject we dealt with in Chap. 1 of the Macropaedia, namely, the relativistic Doppler shift. We want to derive it in a different way that leads to a somewhat more general version of Eq. 1.4. We imagine that there is a frame of reference with respect to which, at the present time, the cosmological background radiation is perfectly isotropic. We are for the moment ignoring the anisotropies induced on the background by the structure-forming events. We also assume that in this frame of reference the photon distribution is blackbody. We are going to ask, What will this distribution look like to an observer moving with a velocity $\mathbf{v}$ with respect to the original frame of reference? It is for this that we need a Doppler shift formula that is valid when the observer is in motion at an arbitrary angle with respect to one of the cosmological background radiation photons. To derive it, we recall the following. The space-time dependence of a plane electromagnetic wave that represents the photon is given by $e^{i[(\mathbf{k}\cdot\mathbf{r})-\omega t]}$, where $\omega/|k| = c$. If we observe this wave in a second frame of reference, moving with a uniform velocity $\mathbf{v}$, the principle of relativity assures us that it will also be a plane wave with an exponential given in terms of $\mathbf{k}'\cdot\mathbf{r}' - \omega' t'$. In fact, we must have

$$\mathbf{k}\cdot\mathbf{r} - \omega t = \mathbf{k}'\cdot\mathbf{r}' - \omega' t' \tag{9.67}$$

since the argument of the exponential is an invariant under the transformations of the special theory of relativity. Figure 5 shows the coordinate transformation that we are interested in.

The relativistic transformation that we need is a generalization of the one we have given where now we allow for the fact that the two coordinate systems are not parallel. We can think of using our old transformation and then rotating the coordinate system. Thus, if we confine things to the xy plane,

$$x = \frac{(x' - vt')\cos\theta}{(1 - v^2/c^2)^{1/2}} + y'\sin\theta \tag{9.68a}$$

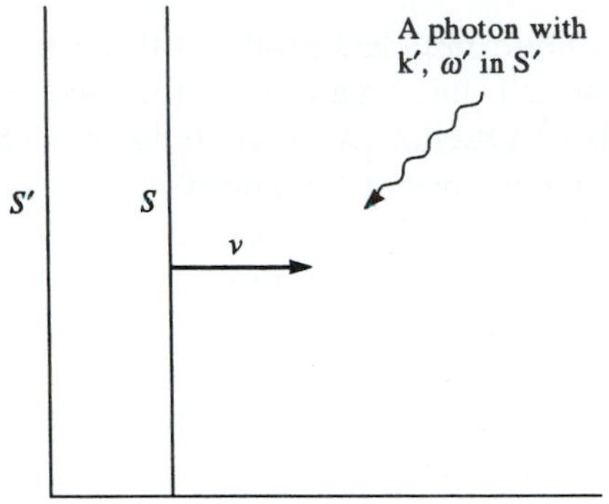

Figure 5 A photon whose wave number and frequency are $\mathbf{k}'$ and ω' in its proper frame is viewed by an observer moving with velocity $\mathbf{v}$.

$$y = y' \cos\theta - \frac{(x' - vt')\sin\theta}{(1 - v^2/c^2)^{1/2}} \tag{9.68b}$$

and

$$t = \frac{t' - v/c^2 x'}{(1 - v^2/c^2)^{1/2}} \tag{9.68c}$$

The reader can verify that these formulas have the right limits and also that

$$x^2 + y^2 - c^2t^2 = x'^2 + y'^2 - c^2t'^2 \tag{9.69}$$

We see from Eq. 9.67 that the four entities $\mathbf{k}$ and ω/c transform like the four entities $\mathbf{r}$ and ct under the transformations given by Eqs. 9.68. This enables us to read off, for example, the transformation of ω, which is just the Doppler shift. Using Eq. 9.68c, we find

$$\omega = \frac{\omega' - (v/c)ck'_x}{(1 - v^2/c^2)^{1/2}} = \omega' \frac{1 - (v/c)\cos\theta}{(1 - v^2/c^2)^{1/2}} \tag{9.70}$$

We see that if $\theta = 0$, we reduce to the results of Chap. 1 of the Macropaedia. If $\theta = \pi/2$, we have $\omega = \omega'/\sqrt{1 - v^2/c^2}$. This is the famous transverse Doppler shift, a consequence of the relativistic time dilatation. It has no counterpart in classical physics according to which the Doppler shift vanishes at right angles.

We now want to use this result to help us find the distortion in the blackbody spectrum as seen by an observer moving with a velocity $\mathbf{v}$ with respect to a system in which the spectrum is perfectly isotropic. The key idea is that the Planck spectral function $1/(e^{pc/kT} - 1)$ takes the same form in all coordinate systems. To argue for this in detail, we would have to go back to the derivation of the Planck spectrum in the relativistic context. This would take us too far afield, but in order not to be too mystifying, let us consider the example of the number density of a particle—say the photon—described by a distribution of the form $f(pc, T)$. We can define a number density current n_α by the equation

$$n_\alpha = \int \frac{d^3p}{(2\pi\hbar c)^3} \frac{p_\alpha c}{E(p)} \times f(pc, T) \tag{9.71}$$

where, for a massless particle, $E = pc$. In the frame of reference in which f is isotropic, all the spatial components of n_α will integrate to zero, leaving only the fourth component, which is the one we identify with the particle density. If $f(pc, T)$ remains invariant under the relativistic transformations, then n_α will transform like a four-dimensional relativistic vector, that is, like $(\mathbf{r}, ct)$. That means, since in the old frame of reference the spatial components of n_α are zero, that the number density—the fourth component—transforms as $n' = n/\sqrt{1 - v^2/c^2}$. But this is just what one wants since, as we recall from our discussion of the Milne model in Chap. 8, n transforms like the total number of particles over the volume. At first sight, this invariance requirement on $f(pc, T)$ seems impossible to meet. Recall that the argument of the exponential in the Planck function is pc/kT. But $pc = \hbar\omega$ and ω transforms according to Eq. 9.70. Thus the only way the argument of the exponential can remain invariant is if the temperature transforms like ω! But this is the result that we are after. That is, we must have

$$T = T'\frac{1 - (v/c)\cos\theta}{(1 - v^2/c^2)^{1/2}} \simeq T'\left(1 - \frac{v}{c}\cos\theta\right) \tag{9.72}$$

where we have used the fact that in practice $v/c \ll 1$. We must be clear how this result is to be applied. We recall from our discussion of the cosmological red shift that there we had a similar kind of invariance. In that case, $p \sim 1/R$ and $T \sim 1/R$, so that pc/kT remained invariant as the universe expanded. But what we did was to focus on a particular p and study the shape of the distribution as the universe expanded. That is how we were able to determine the present temperature of the cosmic background radiation. We are going to do a similar thing here. We will focus on a particular p and allow our radio telescopes to scan the sky at various angles θ. We will then see if there is any variation in the blackbody temperature as a function of the angle θ. Why would we think there might be? It could be that our entire galaxy was in motion with respect to the cosmic background radiation owing to the gravitational attraction of some astronomical object or objects—an interesting idea. In making our angular temperature measurements we remember that θ is the angle between a photon that is incoming to our detector and our velocity relative to the stationary microwave background. With this convention, photons that hit us head-on have a θ of 180°. Their apparent temperature is raised by a factor of $1 + v/c$, while the photons that are chasing us from behind have their temperatures lowered by a factor of $1 - v/c$. We would not have brought all this up if this effect had not been observed. It is called the "dipole anisotropy," since the $1 - v/c \times \cos\theta$ distribution is a typical dipole form. Higher powers of v/c are accompanied by higher powers of the cosine. For example, the coefficient of $(v/c)^2$ would involve $\cos\theta^2$, a quadrupole. The observed dipole effect is a relatively small one. Let us define $\delta T(\theta)/T$ as $\delta T(\theta)/T = |T(\theta) - T|/T \simeq v/c \times |\cos\theta|$. Here T is the mean temperature of the blackbody radiation. For this definition to be useful the fluctuations must be small, so that it is sensible to compare them with the 2.74-K mean temperature. At the most favorable angles, the observations yield $\delta T/T \simeq 10^{-3}$, corresponding to $v \simeq 300 \text{ km s}^{-1}$. This is the speed of the Earth through the stationary microwave background. But the earth also moves with respect to the Milky Way galaxy. Putting these two motions together, observers conclude that the galaxy has a speed of about 600 km s^{-1} with respect to the background. The quadrupole corrections to the dipole are then of the order $(v/c)^2 \simeq 10^{-6}$. The motion seems to be in the general direction of the constellation Hydra. It is natural to suppose that the Milky Way is being pulled gravitationally in this direction. In that case, one should be able to account for the motion by mapping out the gravitational masses—the galaxies or galaxy clusters—that produce it. This is something that astronomers who work with the IRAS have been doing for several years. It seems as if one can actually pinpoint the mass concentrations that are responsible for the dipole motion. One important corrollary to this work is that it gives one a new measurement of Ω since the mass densities of the galaxies out to 500 million light-years have been measured. These measurements produce an Ω which is close to 1—the Holy Grail! It remains to be seen if they stand up to further analysis.

This dipole variation in the isotropy of the background radiation is not what one would call "cosmological" in that there is no apparent early universe explanation of it. It falls into the category of explaining the existence of tigers. In fact, it complicates the study of the cosmological anisotropies and must be subtracted out. Before turning to the *cosmological* anisotropies, we will discuss another example of such an effect which is also not cosmological. It is named after the Russian scientists R. A. Sunyaev and Ya. B. Zel'dovich, who first discussed it in 1972. The idea is that in galactic clusters there can be dilute concentrations—about 10^{-3} cm^{-3}—of electrons that are much much hotter than the ambient cosmic radiation background. These electrons can have average temperatures of as much as 10^8 K. They can scatter elastically from the ambient photons; i.e., there is the reaction $e + \gamma \rightarrow e + \gamma$, which is usually called "inverse Compton scattering." Since these electrons have so much more energy than the cosmic background photons, it is clear that they will heat them up. But these inverse Compton scatterings are "elastic," which means that in such a reaction the number of photons does not change. This means that there will be an inevitable distortion in the blackbody spectrum since, for a blackbody, when one increases the temperature, the number of photons will increase with the cube of the temperature, whereas here there is no increase in the number of photons. What happens is that photons in the low-frequency part of the blackbody spectrum are shifted toward higher frequencies in a specific way that is characteristic of this process and can be computed theoretically. The effect has been observed when radio telescopes are swept across certain galactic clusters. One can measure the distortion in the blackbody spectrum. It is an interesting phenomenon, and it can tell us a good deal about such ionized gasses, but again it is not an early universe effect.

The search for early universe distortions in the cosmic blackbody spectrum has been going on for many years. Up until April 1992, when the COBE satellite reported its results, none had been definitively observed. Since it seemed clear that such effects must be there at some level if the Big Bang theory was to account for structure formation, there was a sense of increasing apprehension in the cosmological community as the level of these effects was being pushed lower and lower by increasingly accurate experiments. Therefore there was a great sense of relief when the COBE result was announced. We shall state the result and then spend the rest of this chapter explaining what it means. One begins by defining a generalization of the quantity $\delta T(\theta)$, the direction-dependent temperature fluctuation we used to discuss the dipole. We may now consider a general direction defined by the angles θ and ϕ. We may define $\delta T(\theta, \phi)$ by the expression $\delta T(\theta, \phi) = T(\theta, \phi) - T$. Like all such angular functions, $\delta T(\theta, \phi)$ can be expanded in the complete set of spherical harmonics $Y_{lm}(\theta, \phi)$; i.e., dividing out by T for convenience, we have

$$\frac{\delta T(\theta, \phi)}{T} = \sum_{lm} a_{lm} Y_{lm}(\theta, \phi)$$

Here the sum runs, at least in principle, from $l = 0$ to $l = \infty$, while for each l, the m sum runs from $-l$ to $+l$. The a_{lm} are angular independent numerical coefficients. The $l = 0$, $m = 0$ coefficient is really included in the definition of the mean temperature T since this term involves Y_{00} which is independent of angle. Hence the sum runs from $l = 1$ to infinity. The $l = 1$ terms are the dipole contributions. These are dropped in these considerations. A better way to put it is that the dipole is measured and then subtracted out of the sum when $\delta T(\theta, \phi)$ is defined. A small part of the dipole arises from intrinsic cosmological fluctuations which contribute, say, a thousandth of the dipole due to the galactic motion, so that they do not interfere with that measurement.

Of course, it would be nice if one could measure the a_{lm} exactly. But this is not possible. One does many measurements and averages over them. Let us denote the averaging procedure, whatever it is, by the symbol $\langle\ \rangle$. By symmetry we must have $\langle a_{lm} \rangle = 0$. What we can hope to measure is $\langle |a_{lm}|^2 \rangle$. We have put the absolute value signs here because the a_{lm} can be complex numbers so long as the sum is real. One can get at the $\langle |a_{lm}|^2 \rangle$ by defining a "correlation function" $C(\mathbf{i} \cdot \mathbf{i}')$, where $\mathbf{i}$ and $\mathbf{i}'$ are the directions defined by θ, ϕ and θ' and ϕ', respectively, as

$$C(\mathbf{i} \cdot \mathbf{i}') \doteq \frac{\langle \delta T(\theta, \phi) \delta T(\theta', \phi') \rangle}{T^2}$$

Using the properties of the Y_{lm} and the averaging procedure, the quantity $C(\mathbf{i}, \mathbf{i}')$ can be enormously simplified. Thus

$$C(\mathbf{i}, \mathbf{i}') = \sum_{l} (2l + 1) C_l P_l(\mathbf{i} \cdot \mathbf{i}')$$

where P_l is the lth Legendre polynomial and we have introduced the angle between the two directions defined by $\mathbf{i}$ and $\mathbf{i}'$ through the scalar product. The sum over l runs between 2 and ∞ if we remove the dipole. Note that after the averaging $C(\mathbf{i}, \mathbf{i}')$ depends only on this single angle. The quantity C_l is defined by the relation

$$\langle a_{lm} a_{l'm'} \rangle = C_l \delta_{ll'} \delta_{mm'}$$

One can in principle pick off any of the C_l's one wants by multiplying $C(\mathbf{i}, \mathbf{i}')$ by the appropriate $P_l(\theta)$ and then integrating over θ. In the language of the C_l's, the dipole is given by $\sqrt{C_1} \simeq 3 \times 10^{-3}$, while the COBE finds $\sqrt{C_2} = 0.76 \pm 0.24 \times 10^{-5}$.

As we have seen, the angular dependence of the temperature fluctuations is of special interest. Fluctuations at angular separations larger than about 1° must come from regions of the sky that were not in causal communication at the time of recombination. Because of the resolution of its telescopes, the COBE experiments cannot probe angular separations of less than about 10°. But for many years measurements of the temperature fluctuations at small angles have been attempted either from balloons or from radio telescopes on the ground. Up until 1993, the results being quoted were upper bounds on these fluctuations. Now several experiments have been reported that actually see small-angle fluctuations at about the level of the COBE. Here the analysis is different. One has two telescopes that point at different angles or one has a single telescope that oscillates back and forth through a small angle. Hence what is compared is the temperature of the blackbody radiation at the small angular separation of the telescopes. Of course, the earth revolves, so that these telescopes sample different portions of the sky over the course of time. Hence there is some averaging. We can define $(\Delta T/T)^2$ by the expression

$$\left(\frac{\Delta T}{T}\right)^2 \doteq \left\langle \frac{[T(\mathbf{i}) - T(\mathbf{j})]^2}{T^2} \right\rangle$$

where $\mathbf{i}$ and $\mathbf{j}$ refer to the directions in which the two telescopes point. But it is not difficult to show—we leave it as an exercise—that we can reexpress this in terms of the $C(\mathbf{i} \cdot \mathbf{j})$ defined above. The relation is

$$\left(\frac{\Delta T}{T}\right)^2 = 2[C(0) - C(\alpha)]$$

where α is the angle between $\mathbf{i}$ and $\mathbf{j}$.

The coefficients—the C_l's that occur here—are the same as those that occur in the COBE experiment since they do not depend on the angle. But there are practical complications. In writing the series expansion for $C(\mathbf{i} \cdot \mathbf{j})$ we have taken the overidealized situation in which the telescope has perfect angular resolution. In practice, each telescope has an angle, call it θ_t, below which it cannot resolve signals. We can take this into account if we introduce a "window function" which mocks up the resolution of the telescope. There is a good deal of arbitrariness as to which function one chooses, but a simple choice that contains the essentials is the Gaussian $\exp(-l^2\theta_t^2/2)$. Here l is the same integer that occurs in the sum over the P_l's. Indeed, we can write with this modification:

$$C(\mathbf{i} \cdot \mathbf{j}) = \sum_l (2l + 1) C_l P_l(\mathbf{i} \cdot \mathbf{j}) \exp\left(-\frac{l^2\theta_t^2}{2}\right)$$

For the COBE, θ_t turns out to be about 0.05 rad. This means that l's larger than about 20 will not contribute to the sum. But it also means that l's less than 20 will. It turns out, and we will go into some of this below, that the most popular theories that attempt to explain why there are fluctuations in the first place also produce connections among the C_l's. In fact, in the simplest such model we have the remarkable relation which holds approximately for small l:

$$C_l = \frac{6C_2}{l(l+1)}$$

This allows one to sum the expression above up to, say $l = 20$, and extract C_2. Since for the Gaussian window function the maximum l is about $1/\theta_t$, in small-angle experiments it is the large values of l that matter. This enables one to test various models by consistency. One can use a formula like the one above to extract C_2 and then use this C_2 to try to fit the small-angle experiments. This should help distinguish among the various models. The results so far are inconclusive. But what do these fluctuations mean?

Until about 300,000 yr after the Big Bang, the cosmic photons scattered so strongly from the electrons that they were essentially trapped. As we have seen, at the time of "recombination"—300,000 yr after the Big Bang—the electrons and protons formed neutral hydrogen. The photons were then released to expand and red-shift freely. The fluctuations in the temperature that we now observe were the fluctuations that were impressed on the photon distribution at the time of recombination. Some cosmologists have entertained the idea that there might be other epochs after recombination in which the neutral matter could reionize. These events, if they exist, would produce additional distortions in the microwave background. As we saw in the beginning of this chapter, any feature in the present cosmos that presents an angular width greater than about 1° must at the time of recombination have come from causally disconnected events. This suggests, as we have been pointed out, that it is useful to divide any presently observed angular fluctuations in the temperature, over and above the dipole, into two classes: those at "small angles" (i.e., less then, say, a degree or so) and those at "large angles." The COBE fluctuations were observed at 10°, so that they are certainly large-angle fluctuations. These large-angle fluctuations then represent features at recombination that are somehow intrinsic to the background radiation since they cannot be produced by microphysical processes at the time of recombination. A photon, for example, cannot have traveled by that time from one side of the fluctuation to the other. These distortions must have been there before recombination. One usually refers to the distribution of particles and radiation at recombination as lying on the last scattering surface. This "surface" is really a volume since it takes a certain time for the recombination to complete itself. But this time is relatively short compared to the 100,000 yr we are talking about, so we will continue to refer to this as a scattering surface. The gravitational potential on this surface might well have irregularities—hills and valleys. A photon climbing out of any of these valleys is red-shifted. If the valleys have different depths, then there will be variations in the red shifts. These will show up as fluctuations in the present microwave background and will be interpreted as fluctuations in the blackbody temperature. This kind of fluctuation is known as the Sachs–Wolfe effect after R. K. Sachs and A. M. Wolf who first discussed it in 1967. It is a serious exercise in general relativity to compute this effect quantitatively, and we shall not try to do so here. It is done in some of the more advanced texts to which we have made reference. Roughly what comes out of this calculation—and this one might expect—is that the fluctuation in the temperature, $\delta T/T$, is proportional to the fluctuation in the gravity-producing matter, $\delta\rho/\rho$. The constant of proportionality is about $\frac{1}{3}$. These fluctuations show up at large angles and presumably are what the COBE has been seeing. An intriguing idea is that they are effects left over from the inflationary epoch. We come back to this later. On the other hand, fluctuations at small angles, in the sense defined above, can have been influenced by microphysical processes, such as electron-photon scattering, that occur at the time of recombination. We have treated recombination, thus far, as if it were something like suddenly opening a window and letting out the trapped light. In reality, the process of liberating the cosmic background radiation is a very complex one. For example, photons are emitted when the electrons are captured by the protons to form atomic hydrogen. These recombination activities must also impress themselves on the cosmic background photons at some level. This sort of distortion should play the dominant role at small angles.

It is a very interesting numerical coincidence that the end of radiation dominance is at about 100,000 yr—the same order of magnitude as the recombination time. This has implications for the temperature fluctuations that we will now explore. It has been known ever since the work done by the British astrophysicist James Jeans in 1902 that the formation of structure—eventually galaxies—involves an interplay between gravity and radiation. In the absence of the pressure supplied by radiation, any gravitational fluctuation would simply grow larger since it would be unopposed and self-enhancing. Newton realized this and, as we have remarked, it led him to conclude that space must be infinite. But if there is radiation pressure, the gravitational fluctuations tend to be smoothed out. Instead of growing, they will develop into oscillating waves, like sound waves. This is what Jeans pointed out. He also pointed out that if the fluctuations were large enough—involved enough mass—there would be gravitational enhancement anyway. Gravitational effects would overwhelm the radiation pressure. He derived a critical crossover length, or mass, which has become known as the Jeans length λ_J or the Jeans mass M_J, at which the effect of radiation pressure and gravitation just balance. We will give a very rough derivation of these quantities that captures the essential physics but does not produce the correct numerical factors, which are of order 1. We do this in the context of the modern expanding universe theory which was of course unknown to Jeans. In the expanding universe, the characteristic time for anything cosmological to happen is $R/\dot{R} \simeq (G\rho)^{-1/2}$, where G is Newton's constant and ρ is the mass density. On the other hand, at the critical mass, or length, we expect the perturbation to propagate with the speed of sound, v_s, characteristic of the material whose mass density we are discussing. Hence it will take a time $t_J = \lambda_J/v_s$ to cross a Jeans length. This means, equating these two times, that the Jeans length is given approximately by $\lambda_J \simeq v_s(G\rho)^{-1/2}$. The Jeans mass of the material involved is given by $M_J = 4\pi/3 \times \lambda_J^3\rho \simeq v_s^3 G^{-3/2}\rho^{-1/2}$. However, the velocity of sound in a material with a pressure

P and a mass density ρ is given by $v_s = (dP/d\rho)^{1/2}$. The discussion now divides itself in two: matter dominance and radiation dominance. In the radiation-dominated regime the pressure is, as we have seen, given by $\rho_E/3$, where ρ_E is the energy density of the radiation. Thus the velocity of sound in such a material is given by $v_s = c/\sqrt{3}$. Since $R/\dot{R} \simeq \tau$, where τ is the time since the Big Bang, we see that the Jeans length in this domain is approximately the size of the entire visible horizon. The fluctuations needed to produce significant matter concentrations would have to be longer than the Jeans length—otherwise they just oscillate and smooth out—which means one would need a fluctuation as large as the visible universe itself at this time.

To make things a bit more quantitative, suppose that there is a perturbation in the matter, which we may call $\delta n/n$. Suppose that this perturbation is "adiabatic," which means that entropy is conserved by it. This means that the total amount of matter, $R^3 n$, is conserved. Therefore in this case $\delta(R^3 n) = 0 = 3R^2 \delta R n + R^3 \delta n$ or $\delta n/n = -3\delta R/R = 3\delta T/T$, where in the last step we have used the adiabaticity to relate R to $1/T$. Thus we see that in this case the matter fluctuations and the radiation temperature fluctuations are tied together. It is the Sachs-Wolfe effect that tells us how fluctuations in the matter density produce corresponding fluctuations in the present photon distribution. We can see a similar effect in a perturbation made under almost the opposite conditions. Suppose the fluctuation takes place on a scale outside the horizon. In that case we may assume that the total energy density ρ remains constant. In general, the total density ρ is the sum of a radiation part and a matter part. We can write, symbolically, $\rho = nmc^2 + AT^4$, where m is the mass of whatever the dominant fluctuating matter is and n is its number density. Here $\delta\rho = 0$ implies $\delta T/T = -mc^2/4 \times n/AT^4 \times \delta n/n$, again a relation, albeit more complex, between matter and radiation fluctuations. In short, if the matter fluctuates, we should expect this to impress itself on the radiation, and if this occurs around the time of decoupling, we should expect to observe it as an anisotropy in the cosmic background radiation. Apparently, this is what the COBE has discovered.

But what is it that fluctuates at recombination? What sort of matter? The most elegant possibility would be the same sort of dark matter that we need to make $\Omega = 1$. To get a feel for this, let us trace through one possibility. Suppose one of the neutrinos, say ν_τ, has a small mass energy, say, to take a round number, 10 eV. What might the consequences be? As we saw in Chap. 7, a neutrino of about 10 eV of the conventional type would close the universe. What sort of a Jeans length, or Jeans mass, would it generate at recombination? Here we must be a little careful. The conventional neutrinos, as we showed, decouple from matter and from each other at equivalent temperatures on the order of a few megaelectronvolts. Thus at the epoch of recombination these neutrinos do not interact with each other. It is therefore incorrect to use the velocity of sound to evaluate their Jeans length. The neutrinos simply pass through each other without leaving a trace. In Chap. 7 we argued (see the discussion leading up to Eq. 7.28) that for a freely expanding neutrino of mass m, $\langle v \rangle/c \simeq kT/mc^2$. We leave it as an exercise for the reader to show that if, instead of the collisionless neutrino, we had considered the proton (which does suffer collisions at recombination) as the fluctuating medium, the corresponding velocity of sound, v_s, would have been given approximately by $v_s/c \simeq \sqrt{kT/mc^2}$. In computing the Jeans length we shall ignore all numerical factors of order 1, including the difference between the photon and the neutrino temperatures. When we write equal signs, we shall mean up to such numerical factors. In this spirit, we write

$$\lambda_J = \langle v \rangle \sqrt{\frac{1}{G\rho}} \tag{9.73}$$

Here ρ is the mass density of the particle of mass m. Introducing the Planck mass M_P and the photon number density n_γ and taking $\rho = nm$, where n is the number density of the neutrinos, we can write

$$\lambda_J = \frac{M_P}{m} \times m\langle v \rangle \times \frac{1}{(\hbar cmn_\gamma)^{1/2}}\left(\frac{n_\gamma}{n}\right)^{1/2} = \frac{\hbar}{mc} \times \frac{M_P}{m}\sqrt{\frac{mc^2}{kT}} \times \sqrt{\frac{n_\gamma}{n}} \tag{9.74}$$

where we have used the expression for $\langle v \rangle$ appropriate to the freely expanding massive neutrino and the fact that $n_\gamma \sim T^3$.

With our approximations, $n_\gamma \simeq n$. Taking a neutrino of 10 eV we have $M_P/m \simeq 10^{27}$ and $\hbar/mc \simeq 10^{-6}$ cm, so that $\lambda_J \simeq 10^{21} \times \sqrt{10\text{ eV}/kT}$ cm. At recombination, $kT \simeq 1$ eV, so that at recombination $\lambda_J \simeq 10^{21}$ cm. But this happens at about 10^{12} s, so that the horizon is at about 10^{22} cm. Hence the Jeans length is within the horizon. If we assume that $\Omega = 1$, then $T \sim 1/\tau^{2/3}$ in this matter-dominated regime so that the Jeans length increases as $t^{1/3}$, while the horizon increases as τ. In the spirit of our approximations, the Jeans mass M_J is given by

$$M_J = \lambda_J^3 mn = m\left(\frac{M_P}{m}\right)^3\left(\frac{kT}{mc^2}\right)^{3/2}\left(\frac{n_\gamma}{n}\right)^{3/2} \tag{9.75}$$

If we put the numbers in and note that the mass of the sun is about 10^{30} kg, we see that at recombination the Jeans mass is in this case about $10^{15}M_\odot$. This is a very suggestive number since galactic clusters have this sort of mass. The picture that it evokes is of a gravitational clumping that eventually leads to the formation of a huge mass—a galactic cluster—that then fragments into individual galaxies. The details are another matter. In fact, there may be a serious, if not fatal, flaw in this scenario—which is an example of what cosmologists call a "hot dark matter" scenario. There is evidence that galaxy cluster formation is going on now. The clusters seem to be forming from the older galaxies. If this is true, then the neutrino scenario may be in serious trouble since it predicts that the galaxies are the result of the breakup of the larger clusters and are therefore younger than the clusters.

The fact that the Jeans mass has this sort of magnitude is characteristic of this scenario. But we must demonstrate that these fluctuations are both plausible and consistent with what we know about the cosmic background radiation. This is something that mut be done to test any model. To approach the latter question we will begin by deriving an equation that governs the time evolution of a perturbation δ in the matter density ρ. The only thing we have to assume about this perturbation is that it preserves the total number of particles. To simplify the analysis we shall also assume that $\Omega = 1$. For the perturbation to preserve the number of particles there must be, as we will now show, a corresponding perturbation in $R_0(t)$, namely, $R = R_0 + \delta R$. Let us define the perturbation in ρ_0, δ, by the equation

$$\rho = \rho_0(1 + \delta) \tag{9.76}$$

We assume that δ is a function of time alone. In general it will be a function of both space and time, but to simplify the discussion we will suppose that the spatial gradients can be neglected. More general perturbations that are functions of space as well as time have been considered, especially in the context of an inflationary origin of the perturbations. We come back to this briefly later. To illustrate what is going on, we now perturb R. The condition of particle conservation is the statement that

$$\frac{d}{d\tau}[(R_0 + \delta R)^3\rho_0(1 + \delta)] = 0 \tag{9.77}$$

If we expand the $R_0 + \delta R$ term and keep only the leading terms in δ and δR, we derive the condition that $d/d\tau\,(\delta + 3\delta R/R_0) = 0$. For simplicity we will take the integration constant to be zero and write

$$\delta = -\frac{3\delta R}{R_0} \tag{9.78}$$

We have used the condition that $d/d\tau\,(R_0^3\rho_0) = 0$, which implies

$$\frac{\dot{\rho}_0}{\rho_0} = -\frac{3\dot{R}_0}{R_0} \tag{9.79}$$

Using these conditions, and we leave this as an exercise for the reader, we can derive the following modified Friedmann equations for R_0 and $R_0 + \delta R$:

$$\frac{d^2}{d\tau^2}R_0 = -\frac{4\pi}{3} \times G\rho R_0 \tag{9.80a}$$

and

$$\frac{d^2}{d^2\tau}(R_0 + \delta R) = -\frac{4\pi}{3} \times G\,[\rho\,(1 + \delta)(R_0 + \delta R)] \tag{9.80b}$$

If we perform the differentiations in Eq. 9.80b and use 9.80a, as well as the identity we get from differentiating 9.78

$$\delta\dot{R}_0 = \frac{\dot{\delta}}{3} \times R_0 - \frac{\delta}{3} \times \dot{R}_0 \tag{9.81}$$

we arrive at the following equation for δ

$$\ddot{\delta} + 2\dot{\delta} \times \frac{\dot{R}_0}{R_0} = 4\pi G\rho_0\delta \tag{9.82}$$

where we have kept only first-order terms in δ and δR. To solve this equation for the time dependence of δ we must make an assumption about the time dependence of R_0. We shall assume that we are in the regime of matter domination so that $R_0(t) \sim \tau^{2/3}$. From the Friedmann equation for R_0 we see that, in this regime,

$$\rho_0 = \frac{1}{6\pi\tau^2 G} \tag{9.83}$$

while $\dot{R}_0/R_0 = 2/3 \times 1/\tau$. Equation 9.83 is consistent with Eq. 9.79, but it is the Friedmann equation that fixed the constants. Thus Eq. 9.82 becomes

$$\ddot{\delta} + \frac{4}{3} \times \dot{\delta} \times \frac{1}{\tau} = \frac{2}{3} \times \frac{1}{\tau^2} \times \delta \tag{9.84}$$

We can solve this equation by letting $\delta = A \times \tau^n$, where A is a constant. If we put this expression into Eq. 9.84, we find a quadratic equation for n, namely, $3n^2 + n - 2 = 0$. This equation has the solutions $n = -1$ and $n = \frac{2}{3}$. The former corresponds to a δ that decreases in time so that no useful structure-forming fluctuation would build up. On the other hand, the $n = \frac{2}{3}$ solution corresponds to a δ that is increasing as $\tau^{2/3}$. Eventually, δ will become of order 1, and the linear approximations that we have been using will break down. Indeed, one of the criteria we can use for the formation of interesting structure is that a useful δ must build up to be 1 no later than the epoch of the present universe. Only when the perturbations become nonlinear does structure formation take place. We will now show that this criterion, adjoined to the time dependence of δ, provides an interesting constraint on any structure-forming model of this kind. Indeed, consider the ratio $\delta(\tau_r)/\delta(\tau_0) = (\tau_r/\tau_0)^{2/3}$, where τ_r is the time at recombination and τ_0 is the present time and where we have used the time dependence of δ that we have just discovered. The ratio $(\tau_r/\tau_0)^{2/3} \simeq 10^{-3}$. This result would seem to rule out protons—ordinary matter—as being the responsible party for galaxy formation. The reasoning runs as follows. At recombination, protons and photons are still in the process of interacting. Hence if the perturbation were protonic, we would expect it to be adiabatic so that $\delta \simeq \delta T/T$. Thus the perturbation in the cosmic background radiation temperature at recombination would be on the order of 10^{-3}. Furthermore, as one can show by computing the Jeans length (we leave this as an exercise), it was well within the horizon, so that these perturbations will now appear at angles of less than 1°. They represent perturbations in the cosmic photons at the time of recombination. Since recombination, the photons have been freely streaming with these fossilized early universe perturbations frozen in. But small-angle perturbations of the cosmic background radiation are now beginning to be observed. They are not as yet known with the accuracy of the large-angle perturbations measured by the COBE, but, nonetheless, it appears that they are all at the 10^{-6} level. These experimental numbers are too small to be consistent with the protonic fluctuations we have just estimated. At least in this respect, neutrino fluctuations appear to be more promising. At recombination neutrinos do not interact with the photons, so any connection they have with the photon fluctuations must have come from an earlier epoch. Such early fluctuations might well have been washed out, so it may be that one is safe from the proton dilemma. On the other hand, there are some unattractive features of the neutrino scenario which may not flatly rule it out but which make it seem somewhat less desirable. In the first place, there is the size of the Jeans mass. It is comparable to that of a galactic cluster. What then accounts for the myriad of smaller mass objects such as ordinary galaxies? This must involve some entirely different kind of physics which then must be supplied. Then there is the age problem alluded to above. Of course, if the tau neutrino actually turns out to have a mass energy of 10 eV or so, one can rest assured that cosmologists will find a way to overcome these difficulties. It is also possible to consider scenarios involving hypothetical very massive (1 GeV or more) neutrino-like objects—weakly interacting massive particles (WIMP)—which have some of the desirable features of the neutrinos but which would produce Jeans masses closer to the galactic scale. These are called cold dark matter scenarios and are the subject of very active research. They would be much more compelling if experiment actually revealed the existence of such particles.

The fluctuation scheme that is attracting most attention at present attempts to identify the fluctuations that are generated during inflation with the fluctuations that cause ripples in the cosmic background radiation. As we have seen, inflationary quantum fluctuations may be produced by one or more unstable scalar mesons that no longer exist, although they could presumably be produced in accelerators. Since these mesons decay at the end of the inflationary period, their effect on the cosmic background is assumed to have been produced during inflation. This has a very nice feature. Let us suppose that these fluctuations have some sort of spatial dependence $\delta\phi(x)$. It is then always possible to expand $\delta\phi(x)$ in a Fourier series or to represent it as a Fourier integral. We can then focus on a particular wave number k or a particular wavelength λ in the expansion. As the universe expands, this wavelength is red-shifted by $R(t)$. During the inflationary epoch $R(t)$ is increasing exponentially. But the quantity $cR/\dot{R}$ is constant. This is what gives rise to the exponential. But $cR/\dot{R}$ is approximately the distance over which a photon can interact during this unfolding of the expansion. Hence the perturbation is inflated exponentially outside the causal horizon during the brief inflationary epoch. After it is over, radiation dominance is restored and $R/\dot{R} \sim t$. Hence the causal horizon will move out to catch up with the perturbation length scale. Figure 6 illustrates the situation. In this picture,

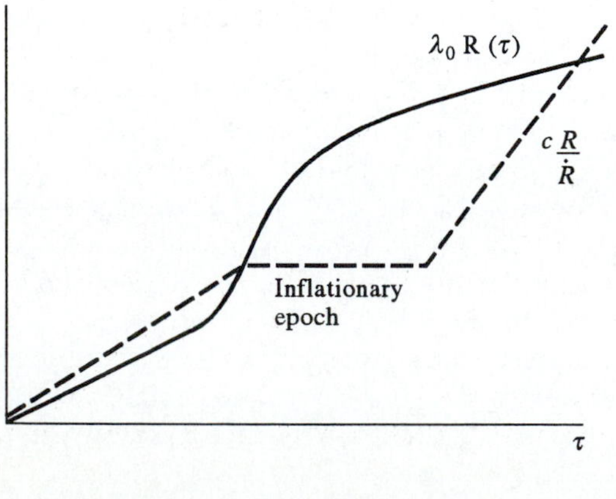

Figure 6 The causal size $Cr/\dot{R}$ and the red-shifted perturbation wavelength plotted against time.

the ripples that the COBE has now discovered are fossilized remains from the preinflationary epoch—about 10^{-35} s after the Big Bang! Some promising models have been explored, but again, "only a few details are missing."

This chapter has taken us to the frontier of our subject. The reader may share the view of the writer that an enormous amount of progress has been made since 1965 when the cosmic background radiation was first discovered. It is also clear that many problems—perhaps the hardest problems—remain to be solved. But this may have to do with the fact that we don't have as yet quite enough observational data, something that one hopes will be remedied in the next few years. So the reader may have the feeling, shared by the writer, that he or she has been left suspended in midair. But a text on cosmology, unlike a novel, not only has no happy ending but has no real ending at all.

Problems

MICROPAEDIA

1.1. Assuming that the mass density of the universe at time t is isotropic, that is, $\rho(\mathbf{r}, t) = \rho(|\mathbf{r}|, t)$, where $\mathbf{r}$ is measured from any point, show that ρ is homogenous, i.e., that ρ is independent of $\mathbf{r}$. Is the converse true? Does homogeneity imply isotropy? If not, can you find a counterexample? Ignore all effects of relativity.

1.2. Using the Fermi constant G_F, $\hbar$, and c, construct a set of "Fermi units" analogous to the Planck units; i.e., find the Fermi mass, the Fermi temperature, and so on. This is an exercise in dimensional analysis.

1.3. Show by using dimensional analysis that a massless particle gas, like the photon, in equilibrium at the temperature T has a number density proportional to T^3 and an energy density proportional to T^4. The players in the dimensional game are T, k, c, and $\hbar$.

1.4. In Yukawa's original weak interaction model the pi meson was supposed to mediate both the weak and strong couplings. The pi meson in this model coupled to the nucleons with a strong coupling g_s and to the leptons with a weak coupling g_w, with $g_s \gg g_w$. Show, by drawing the relevant Feynman diagrams, that this model does not allow for the "universality" of the weak interactions. The processes $\nu + p \rightarrow e^+ + n$ and $\nu + e^+ \rightarrow \nu + e^+$ would, contrary to experience, take place at very different rates. The model of the weak interactions that we use in cosmology, on the other hand, has universality built into it. Explain.

MACROPAEDIA

Chapter 1

1.1. The Lorentz transformations that connect the systems S and S′ of Fig. 1 are given by

$$x' = \frac{x - vt}{(1 - v^2/c^2)^{1/2}}$$

$$y' = y$$

$$z' = z$$

$$t' = \frac{t - v/c^2\, x}{(1 - v^2/c^2)^{1/2}}$$

An "event" in relativity is specified by four coordinates x, y, z, ct, where we use ct rather than t for dimensional homogeneity. The "distance" between two events Δs^2 is given by

$$\Delta s^2 = (x_1 - x_2)^2 + (y_1 - y_2)^2 + (z_1 - z_2)^2 - c^2(t_1 - t_2)^2$$

Show that Δs^2 is the same whether expressed in the S or S′ coordinates. Δs^2 is what is known as a "Lorentz scalar" quantity. Some authors use the convention $-, -, -, +$ for the signs in the "metric" defined by the Δs^2, rather than the convention $+, +, +, -$ that we have used. Either convention will do so long as one sticks with one's choice.

1.2. One can derive the Lorentz transformations from the requirement that they leave Δs^2 invariant. This exercise will show how to derive the Lorentz contraction and the time dilatation from the invariance of Δs^2. First the time dilatation.

(a) The "proper time" is the time that any observer at rest with respect to a clock measures. Each observer measures the same proper time, so that it is a Lorentz scalar. Indeed, the square of the proper time interval between two events is defined to be $\Delta\tau^2 = -\Delta s^2/c^2$. To find $\Delta\tau$ we take the sign of the square root that gives the correct time ordering. In what follows we will consider only motions along the x axis so that y and z do not enter the problem. The two events to be considered are the coincidences of the origins of S and S′ at $t = t' = 0$ and the arrival of the origin of S′ at a new location along the x axis of S a time t later according to S. How much proper time has elapsed according to S′, and how is this time related to t?

(b) To determine the Lorentz contraction we construct the Δs^2 that connects the following two sets of events. At $t = t' = 0$ the origins of S and S′ coincide. At a later time t, as measured in S, the origin passes the point L, as measured in S. This is the second event. Using Δs^2, find the distance L', as measured in S′, that corresponds to L.

1.3. (a) Two reference frames S and S′ are oriented so that S′ is rotated through an angle θ in the xy plane. S′ is moving with a uniform velocity whose components, when measured in S are

$$v_x = v \cos\theta$$

$$v_y = v \sin\theta$$

Find the Lorentz transformation between these frames. *Hint*: Make a Lorentz transformation followed by a rotation. We will need this result in Chap. 9.

(b) Show that your results satisfy

$$x^2 + y^2 - c^2t^2 = x'^2 + y'^2 - c^2t'^2$$

1.4. Derive the relativistic Doppler shift formula for the case in which the observer is moving away from the source. Note that in classical physics the two situations—source at rest, observer moving, and observer at rest, source moving—are not equivalent. What differences are there between the Doppler shift for light and the Doppler shift for sound from the point of view of relativity?

1.5. Using nothing more complicated than the Pythagorean theorem and some simple facts about relativity, show that there is a "transverse" Doppler shift for light; i.e., when the observer is moving at right angles to the source, the wavelength seen by the observer is Doppler-shifted. How big is this effect for the transverse motion of stars as observed in a typical galaxy? We will rederive this result in Chap. 9 in a more sophisticated way.

1.6. Using Hubble's law and assuming that $h = 1$, how many kilometers a year does the Hyades cluster recede from us? How does this number compare to the radius of the earth?

Chapter 2

2.1. Many books on cosmology introduce a quantity called the "deceleration parameter" q_0. The subscript zero refers to the present time. It has the definition

$$q_0 \doteq -\ddot{R}(t_0)\frac{R(t_0)}{\dot{R}(t_0)^2}$$

(a) Using the various Friedmann equations, show that for a matter-dominated model,

$$q_0 = \tfrac{1}{2}\Omega_0$$

(b) Show that

$$Z = \frac{R(t_0)}{R(t)} - 1 = H_0(t_0 - t) + (1 + \tfrac{1}{2}q_0)H_0^2(t_0 - t)^2 + \ldots$$

where H_0 is the Hubble constant.

(c) Find the corresponding expansion for $t_0 - t$.

2.2. General relativity provides a generalization of Eq. 2.12, namely,

$$\frac{\ddot{R}}{R} = -\frac{4\pi G}{3}(\rho + 3P)$$

where P is the pressure. Matter domination corresponds to the case of negligible pressure. In all cases, the Friedmann equation for $\dot{R}/R$ that we used in the text is correct provided that we are careful to use the mass density or its equivalent. Find the deceleration parameter in this more general case in terms of Ω_0, P, and ρ. What is q_0 for the case of radiation dominance?

2.3. How many protons per cubic centimeter would have to be created in a year in order to maintain the proton number density at its present value in light of the expansion of the universe? If that were arranged and the universe were proton matter-dominated, how would it expand?; i.e., find $R(t)$.

2.4. Evaluate the integral in Eq. 2.62 for the case in which $0 \leqslant \Omega \leqslant 1$. Check the two limiting cases $\Omega = 0, 1$. Find the present time t_0 as a function of Ω.

Chapter 3

3.1. A common situation is one in which a material target is irradiated in, say, a cyclotron. We shall suppose that the target consists of N_A stable nuclei of type A. These nuclei are transformed in part into radioactive nuclei of type B that were not present before the machine was turned on. The machine produces a "flux" of irradiating particles that we call $\mathscr{F}$. This is the number of projectiles from the cyclotron that strike each square centimeter of the target in each second. Since there are so many more stable nuclei in the target as compared to the number in the beam, we shall treat N_A as if it were constant in time during the irradiation. Hence we can write the constant rate at which the B nuclei are produced as $R = N_A\,\sigma\mathscr{F}$. Here σ is a constant of proportionality called the "cross section". Given that the decay rate of B is λ_B, find an equation for $N_B(t)$ and solve it. The "activity" of the sample—in this case the target—is defined to be $A(t) = \lambda_B N_B(t)$.

Assuming that the half-life of B is much longer than the time the cyclotron is kept running, find an approximate expression for $A(t)$.

3.2. Climbers on Mount Ararat in Turkey discover wooden timbers which they think may be the remnants of Noah's ark. They find that the ^{14}C component of the wood gives three decays per minute. At base camp they find a tree with the same wood content that gives six decays per minute. What is the age of "Noah's ark?"

3.3. In 1801, a German mathematician and surveyor named Johann Georg von Soldner published a paper in which he predicted that light would be bent by a gravitational field. His idea was that light consisted of particles with a mass m, and he then applied Newton's law to compute their bending. Let us use these ideas to find the condition relating the radius of a gravitating sphere to its mass so that this sphere is a black hole. In other words, light emitted by an object at the surface of this sphere cannot escape. In doing this exercise use Newtonian mechanics and not relativity. Surprisingly, the condition that emerges gives what is known as the "Schwarzchild radius" for a black hole in general relativity. Find this radius for the earth, for the sun, and for a proton.

Chapter 4

4.1. In the nineteenth century, James Clerk Maxwell derived the formula that bears his name, which gives the average number of molecules per unit volume of mass m in equilibrium at temperature T with speeds between v and $v + dv$:

$$F(v)\,dv = 4\pi n\left(\frac{m}{2\pi kT}\right)^{3/2} \times v^2 e^{-mv^2/kT}\,dv$$

Here n is the density of molecules. When we compute averages, we divide by n or, equivalently, we can normalize F so that

$$\int_0^\infty F(v)\,dv = 1.$$

(a) Normalize F.
(b) Using this normalized F, compute $\langle v \rangle$, where

$$\langle v \rangle = \int_0^\infty vF(v)\,dv.$$

(c) Compute $\langle v^2 \rangle$ and show how this differs from $\langle v \rangle^2$.
(d) Use F to compute the most probable value of v.

4.2. Use dimensional analysis to find Eq. 4.6.

4.3. Consider a gas made up of massless particles in equilibrium at temperature T. Suppose that this gas has no chemical potential. Compute the n_i of Eq. 4.43 for both the Bose–Einstein and Fermi–Dirac cases and take the ratio. We will need this number in Chap. 7. These integrals are not doable in terms of what one usually calls elementary functions, but in this day and age of computers that should be no problem.

4.4. Take the logarithm of both sides of Eq. 4.61 and solve for T_c, including the $T^{3/2}$ term. Plot X_p as a function of T, taking $\eta = 10^{-9}$, and be amazed by how dramatically the ionization sets in and how rapidly it is completed.

Chapter 5

5.1. In the so-called Sakata model, the "quarks" are the particles p, n, and Λ^0. One then tries to construct everything else out of them. Baryons are composed of two "quarks" and an "antiquark." Can you construct all the baryons in Fig. 4 using these "quarks"? How does the quark model avoid any problems you may run into when you use the "quark" model?; i.e., what quirk does the quark model have that the "quark" model doesn't?

5.2. The mean free path d between interactions, with a cross section σ of a particle moving through a density of target particles, n, is defined as $d = 1/\sigma n$; that is, the larger the cross section and/or the density, the shorter the mean free path. Imagine a neutrino with energy E moving through lead. Then,

(a) Using dimensional analysis find an expression for σ in terms of the Fermi constant, $\hbar$, c, and E. Exercise 2 of Chap. 4 is a good warm-up. The atomic number of lead is 82, and the density is $11.34\ \text{g cm}^{-3}$. Ignore the binding energy of lead and the neutron proton mass difference and find the mean free path of a 1-GeV neutrino in lead.

(b) Suppose you had a material with the same n made up of only electrons. What would the mean free path of a photon be in this material if its cross section were simply r_0^2, where r_0 is the classical electron radius. In reality, the photon can have all sorts of interactions with atomic electrons. Name some.

5.3. What conservation laws, or symmetries, forbid the following reactions:

$$\pi^0 \rightarrow \gamma + \gamma + \gamma$$

$$\mu^+ \rightarrow e^- + e^+ + e^+$$

$$\bar{p} \rightarrow e^- + \nu_e + \bar{n}$$

$$\gamma + e^- \rightarrow e^-$$

$$p \rightarrow \pi^0 + e^+$$

$$\pi^- + p \rightarrow \Sigma^0 + \bar{K}^0$$

$$\nu_e + p \rightarrow e^+ + \Lambda^0 + K^0$$

Chapter 6

6.1. Using only two-body collisions and strong interactions, find the sequences of collisions that can produce ^{4}He starting with deuterium and nucleons.

6.2. Show, by doing the integrals in Eq. 6.31, that the critical temperature at which electron-positron annihilation takes place is given by $kT \simeq m_e c^2/5$. How sensitive is this number to the criterion we have used? Take various values of the right-hand side of Eq. 6.31 and see.

6.3. At about $kT = 100$ MeV, μ^+ and μ^- annihilate. What is the rise in the photon temperature after the annihilation, assuming entropy conservation? For purposes of the problem, neglect the effect of the muon's instability. What is the rise in the neutrino temperature?

6.4. Include the effect of the $(m_p c^2/kT^\gamma)^{3/2}$ term in Eq. 6.49 and determine the temperature and the time at which deuteron formation takes place.

6.5. Solve for T^γ in Eq. 6.51.

6.6. Fill in the details of the derivation of Eq. 6.74.

6.7. Determine the magnitude of μ_{ν_e}/kT needed to compensate for an additional neutrino flavor in helium production.

6.8. Given that the binding energy of ^7Li is about 39.3 MeV, use the equilibrium approximation and Eq. 6.106 to find the temperature at which ^7Li would be formed. Is this what happens? In working this out, note that ^7Li has an intrinsic angular momentum of $\frac{3}{2}$ in $\hbar$ units.

Chapter 7

7.1. Derive the result in Eq. 7.1. You will find using the relativistic four-vector notation a true blessing. If one neglects all the recoil momenta and the electron and neutrino masses, a much simpler formula can be derived nonrelativistically. Do this as a warm-up exercise and show how Eq. 7.1 reduces to this formula in the appropriate limit.

7.2. Study a table of nuclides and show that the threshold energy for the reaction $\nu_e + {}^{37}Cl \rightarrow e^- + {}^{37}Ar$ is 0.86 MeV.

7.3. Derive Eq. 7.3 and find out how well you would have to know the momentum to measure m_{ν_μ} to 1 percent.

7.4. Using the ideas leading to Eq. 7.8, find $P_{\nu_e\rightarrow\nu_e}(t)$ and verify that $P_{\nu_e\rightarrow\nu_\tau}(t) + P_{\nu_e\rightarrow\nu_e}(t) = 1$ at all times t.

7.5. Verify Eq. 7.10.

7.6. Using the definition of $\langle p \rangle$ show that, for a massless neutrino, $c\langle p \rangle = 3\,kT$.

7.7. Derive Eq. 7.33.

Chapter 8

8.1. Show that for an isotropic universe there are no $dx\,dy$ terms in the metric.

8.2. Prove Eq. 8.12.

8.3. Suppose the conditions of Problem 3 of Chap. 2 had been maintained from $t = -\infty$ to the present. This is the so-called Steady State theory, which had a certain vogue before the cosmic background radiation was discovered. What would the event horizon and particle horizons be in this model?

8.4. Derive Eqs. 8.39 and 8.40.

Chapter 9

9.1. Suppose Ω_0 should turn out to be, say, $\frac{1}{2}$. Assuming the Standard Model, how close to 1 would Ω have to be, at $kT = 10^{14}$ GeV, for this to be realized. Make the adjustment, when appropriate, from the matter-dominated to the radiation-dominated regime.

9.2. Verify Eqs. 9.21 and 9.24.

9.3. Introducing a cosmological constant opens up new possibilities for model universes which probably don't have much to do with reality but are fun to contemplate. We approach this matter as follows:

(a) Using Eq. 9.31, show that if Λ is less than or equal to zero, the universe begins from an initial condition, which we can write choosing the initial time to be zero as $R(0) = 0$.

(b) If Λ is positive, then there is a new possibility we can explore. In this case, $R(t)$ is no longer convex away from the t axis. In fact, there can be an instant at which $\dot{R} = 0$, provided that $k = 1$. This can correspond to a nonzero minimum R. In this circumstance, the universe would never pass through a singular initial state. To be more specific, let us assume that this model universe is matter-dominated throughout its history. In that case, Eq. 9.32 becomes a cubic equation that determines, when it has a real solution, the value of R at which $\dot{R} = 0$. Show that there is such a solution when Λ satisfies the inequality $\Lambda < c^6/(4\pi G\rho_0 R_0^3)^2$. Here the subscripts 0 refer to the present time.

(c) If you set Λ equal to this quantity and use the present proton density, how large is the ρ_V corresponding to this Λ as compared to the critical density with $h = 1$?

9.4. Verify Eq. 9.40 and the value of ϵ given in the text.

9.5. Use Eq. 9.58 and the discussion following it to fix the value of λ.

9.6. Verify Eq. 9.75 and compute the Jeans mass for a 10-eV neutrino.

9.7. Find $\delta(\tau)$ in the radiation-dominated epoch. Begin by showing that the relevant equation for δ in this epoch is

$$\ddot{\delta} + 2\frac{\dot{R}}{R} \times \dot{\delta} = \frac{32\pi}{3} G \times \rho_0$$

and using the Friedmann equation here to find ρ_0 as a function of τ.

9.8. The purpose of this lengthy problem is to acquaint the reader with the essentials of the so-called Newtonian theory of fluctuations of a gravitating medium and its effect on the observed blackbody spectrum. The methods used here can be justified by the general theory of relativity, and the reader is invited to consult one of the more advanced books cited in the references for such a treatment. This problem will make more quantitative some of the discussion in the last part of Chap. 9. The subproblems have been arranged so as to lead the reader through this development in a stepwise fashion.

(a) In Chap. 1 of the Macropaedia we discussed the red shift of a photon in a uniform gravitation field and arrived at the formula Eq. 1.25:

$$\frac{\delta\nu}{\nu} = \frac{\nu' - \nu}{\nu} = -\frac{gd}{c^2}$$

where g is the gravitational acceleration, d the height above some ground level, and c the speed of light. Show that this is a special case of the formula

$$\frac{\delta\nu}{\nu} = \frac{\phi}{c^2}$$

where ϕ is the gravitational potential per unit mass. Derive this more general formula.

(b) Recall that the potential ϕ satisfies the equation

$$\nabla^2\phi = 4\pi G\rho$$

where ρ is the mass density and G is the gravitational constant. We are now going to perturb ρ about a spatially constant mass density $\rho_0(\tau)$. We write this perturbation in the form (see Eq. 9.76)

$$\rho(\mathbf{r}, \tau) = \rho_0(\tau)[1 + \delta(\mathbf{r}, \tau)]$$

where $\delta(\mathbf{r}, \tau)$ is the perturbation. Show that if we write

$$\phi(\mathbf{r}, \tau) = \phi_0(\mathbf{r}, \tau) + \delta\phi(\mathbf{r}, \tau)$$

then $\delta\phi$ obeys the equation

$$\nabla^2\delta\phi(\mathbf{r}, \tau) = 4\pi G \times \rho_0(\tau) \times \delta(\mathbf{r}, \tau)$$

with the solution

$$\delta\phi(\mathbf{r}, \tau) = -\rho_0(\tau)G \times \int d^3r' \frac{\delta(\mathbf{r}', \tau)}{|\mathbf{r} - \mathbf{r}'|}$$

All of this has been done in flat nonexpanding space. We can take the expansion into account if we scale the r coordinates by the scale factor $R(\tau)$, that is, let $\mathbf{r} \rightarrow R(\tau)\mathbf{r}$, where τ is now to be thought of as the proper time. Thus, written in terms of these coordinates, the above becomes

$$\delta\phi(\mathbf{r}, \tau) = -\rho_0(\tau)G \times R^2(\tau) \int d^3r' \frac{\delta(\mathbf{r}', \tau)}{|\mathbf{r} - \mathbf{r}'|}$$

(c) Let us assume that $\delta(\mathbf{r}, \tau)$ can be written as

$$\delta(\mathbf{r}, \tau) = \delta(\tau) \times f(\mathbf{r})$$

In Chap. 9 we derived an equation for $\delta(\tau)$ under the assumption that $f(\mathbf{r})$ was slowly varying so that we could neglect spatial gradients. Here we need this equation under the additional assumption of matter dominance (Eq. 9.84). The reason is that the perturbations we observe in the cosmic background radiation, barring reionization, were the ones imposed on it at the time of recombination, which coincided with the onset of matter dominance. Under these conditions we found that the equation had a growing solution of the form $\delta(\tau) \sim \tau^{2/3} \sim R(\tau)$ since in the regime of matter dominance, we have $R(\tau) \sim \tau^{2/3}$. We will absorb the constant of proportionality in $f(\mathbf{r})$ and write

$$\delta\phi(\mathbf{r}, \tau) = -\rho_0(\tau)G \times R(\tau)^3 \int d^3r' \frac{f(\mathbf{r}')}{|\mathbf{r} - \mathbf{r}'|}$$

But in this matter-dominated regime,

$$\rho_0(\tau)R^3(\tau) = M$$

where M is the total mass, which we assume to be constant in time. This means that $\delta\phi(\tau)$ is, in the regime of interest, independent of τ. Assuming that $\Omega = 1$ and using

$$H^2 = \left(\frac{\dot{R}}{R}\right)^2 = \frac{8\pi}{3} \times G\rho_0$$

we can rewrite the expression for $\delta\phi$ as

$$\delta\phi(\mathbf{r}, \tau) = -\frac{3}{8\pi} H^2R^2 \int d^3r' \frac{\delta(\mathbf{r}', \tau)}{|\mathbf{r} - \mathbf{r}'|} = -\frac{3}{8\pi} H_0^2 R_0^2 \int d^3r' \frac{\delta(\mathbf{r}', \tau_0)}{|\mathbf{r} - \mathbf{r}'|}$$

where we have taken advantage of the time independence of $\delta\phi$.

To get a feeling of what is going on, suppose that at some time τ there is a perturbation that is nonzero and constant over some finite region in r space. We can then write a rough approximation to $\delta\phi$ as

$$\delta\phi \simeq -H(\tau)^2 \times [R(\tau)r]^2\, \delta(\tau)_r$$

where we have ignored the various constants of order 1. Here $\delta(\tau)_r$ is a dimensionless number that sets the magnitude of the perturbation of scales r. We have from experiment a value of $\Delta T/T \simeq 10^{-5} \simeq \delta v/v$. Use the expression above and this present-day data to set a limit on the size of $\delta(\tau)_r$ at recombination if we assume that $R(\tau)r$ is the size of the causal horizon at recombination. Can the COBE with its 10° resolution measure any perturbations on a length scale smaller than this?

References

INTRODUCTION

There is as yet no definitive history of modern cosmology. J. Bernstein and G. Feinberg, *Cosmological Constants*, New York, Columbia University Press, 1986, collect some of the historically important papers with commentary. In *The First Three Minutes*, Basic Books, New York, 1977, S. Weinberg presents some interesting historical material.

MICROPAEDIA

The pot pourri of material in this chapter is derived from various sources. The physical constants and particle properties are taken from *Physical Review D*, 45, 11, 1992. A useful general astronomy text at something like the level of this one is *Astrophysical Concepts* by M. Harwit, New York, Springer, 1988. Statistical mechanics plays a crucial role in our subject. A solid text is *Statistical and Thermal Physics*, by F. Reif, New York, McGraw-Hill, 1965. An overview on a semipopular level is given by J. Bernstein in *The Tenth Dimension*, New York, McGraw-Hill, 1989.

Chapter 1 The definitive text on astronomical distances is *The Cosmological Distance Ladder* by M. Rowan-Robinson, New York, W. H. Freeman, 1985. For relativity there is the fine text *Essential Relativity* by W. Rindler, New York, Springer, 1979.

Chapter 2 There are many books on theoretical cosmology. The closest in spirit and scale to this one is *Principles of Cosmology and Gravitation* by M. V. Berry, New York, Adam Hilger, 1976. I have given the original date of publication because the book has not been revised although there is a second edition. Berry focuses more on the formal theory than I do, but what he covers he does so with wonderful clarity. However, a lot has happened since 1976. The mother of all cosmology books is *Gravitation and Cosmology* by S. Weinberg, New York, John Wiley, 1972, but it is even older than Berry's book, and Weinberg has not chosen to update it. On a level with Weinberg I would put P. J. E. Peebles's *Physical Cosmology*. Happily, Peebles updated this book in 1993, *Physical Cosmology*, Princeton, N.J., Princeton University Press, 1993. Both of these books are for a more advanced reader than this one is meant to serve. If this book fulfills its purpose, readers will be able to tackle the more advanced texts after studying it. There are several newer books. Among them I have found useful *The Early Universe* by G. Börner, New York, Springer, 1988, and *The Early Universe* by E. W. Kolb and M. S. Turner, Redwood City, CA, Addison Wesley, 1990. A very charming short book with interesting insights is *Creation of the Universe* by Fang Li Zhi and Li Shu Xian, Teaneck, N.J., World Scientific, 1989. I would also recommend a new text by T. Padmanabhan, *Structure Formation in the Universe*, Cambridge University Press, New York, 1993. It is also more advanced than this one.

Chapter 3 A solid nuclear physics text is *Introductory Nuclear Physics* by K. S. Krane, New York, John Wiley, 1987.

Chapter 4 Reif, op. cit., and the references for Chap. 2. See also *Kinetic Theory in the Expanding Universe* by J. Bernstein, New York, Cambridge University Press, 1988.

Chapter 5 In addition to my book *The Tenth Dimension*, already cited, on a much more advanced level is *Cosmology and Particle Physics* by R. Domínguez and M. Quirós, Teaneck, N.J., World Scientific, 1988.

Chapter 6 The nonequilibrium helium production model cited in the text can be found in "Cosmological Helium Production Simplified" by J. Bernstein, L. S. Brown, and G. Feinberg, *Reviews of Modern Physics*, 61, 1, 1989. See also my book on kinetic theory cited above.

Chapter 7 A very useful reference is *Physics of Massive Neutrinos* by Felix Boehm and Peter Vogel, New York, Cambridge University Press, 1992.

Chapter 8 The use of the Milne model as a teaching tool was suggested to me by Rindler's book on relativity cited above which also has an excellent discussion of horizons.

Chapter 9 The more recent cosmology books cited above all have material germain to this chapter. In addition I have found *Development of Large-Scale Structure in the Universe* by J. P. Ostriker, Pisa, Pantograf, 1991, very useful. A serious reader can study *Large Scale Structure in the Universe* by P. J. E. Peebles, Princeton, N.J., Princeton University Press, 1980. See also the daily newspapers.

Index